Measurements of Multi Paced Solutions By Processed Quaternary Chalcogenides

T. Laxmi

TABLE OF CONTENTS

LIST OF FIGURES

LIST OF TABLES

LIST OF SYMBOL AND ABBREVIATIONS

CFTS	-	Cu_2FeSnS_4
CFBTS	-	$Cu_2Fe_{1-x}Ba_xSnS_4$
AM1.5	-	Air mass 1.5
θ	-	Angle of diffraction
e	-	Charge of electron
D	-	Crystallite size
J-V	-	Current density– Voltage characteristics
η	-	Efficiency
E_g	-	Energy band gap
EDAX	-	Energy dispersive X- ray analysis
FF	-	Fill factor
β	-	Full width at half maximum
FWHM	-	Full width at half maximum
HRTEM	-	High resolution transmission electron microscope
P_{in}	-	Incident light intensity
ITO	-	Indium doped tin oxide
P_m	-	Maximum power
FTO	-	Flourine doped tin oxide
nm	-	nano meter
V_{oc}	-	Open circuit voltage
h	-	Planck's constant

SEM	-	Scanning electron microscope
SAED	-	Selected area electron diffraction
kWh	-	kilo Watt hour
mL	-	Milliliter
µL	-	Microliter
nm	-	Nanometer
sq. m	-	Square meter
PV	-	Photovoltaic
c-Si	-	Crystalline silicon
CdTe	-	Cadmium telluride
PCE	-	Power conversion efficiency
TW	-	Terawatt
eV	-	Electron volt
δ	-	dislocation density
E	-	Micro strain
n	-	refractive index
ε_∞	-	dielectric constant
I-t	-	Current Vs. time characteristics
XPS	-	X-ray photoelectron spectroscopy
XRD	-	X-ray diffraction

Abstract

Our research focuses mainly on the development of new techniques and materials for the fabrication of low-cost thin film light absorbing layers and their multifunctional applications. The development of new materials for energy conversion and storage has broadened their utilization into areas where their semiconducting and conducting properties have encouraged use in many novel applications. There are several semiconducting materials that can be used to fabricate an absorber layer, among those the development of quaternary earth abundant chalcogenide compounds such as Cu_2ZnSnS_4 (CZTS) was initially considered very promising compounds with efficiency 12.6 %. This efficiency is not exceeded till date due to some of the defects and secondary phases. The different types of defects are the point defects which include vacancies, antisites and interstitial defects, and acceptor and donor defects. Cu_{Zn} anitisite is the lowest formation energy acceptor in CZTS. Whereas, the other acceptor defects like V_{Cu}, V_{Zn}, Zn_{Sn} and Cu_{Sn} have higher formation energy than the dominant Cu_{Zn} antisite defect. Also the donor-acceptor pair transitions, spatially fluctuating band states and band-to-impurity transitions are also reported. It is difficult to synthesis pure CZTS without any secondary phases due to these defects and thus forms binary and ternary secondary phases. As the stability of the CZTS material is comparatively less and the biggest challenge is to develop a phase pure structure without any secondary phases. Also the reduction in the antisite defects by cation substitution has received considerable attention. The presence of Cu/Zn cation disorder causes more detriminal to the electronic transport in the CZTS. Therefore the replacement of Zn with other metals like Fe, Ba, etc could help to reduce these defects.

In this thesis, as a potential replacement, Zn from the quaternary chalcogenide based semiconductor CZTS was replaced by a transition metal known as iron (Fe) to form Cu_2FeSnS_4 (CFTS). As they are earth-abundant and non-toxic, having an optimal band gap in the range of 1.0-1.5 eV with higher absorption coefficient ($>10^4$ cm^{-1}) in the visible spectrum range makes them well suited for multifunctional applications. A solution based approach was adopted for the fabrication of p-type CFTS absorber material. We adopted three different methods for the synthesis of CFTS nanoparticles which includes hydrothermal, sol-gel and successive ionic layer adsorption and reaction (SILAR) methods.

We have successfully synthesized different types of CFTS nanoparticles with the help of solvothermal route using PEG with different chain length as solvent followed by sulfurization. The crystallinity of the CFTS nanoparticles drastically increased after sulfurization which was confirmed by X-ray diffraction (XRD) and Raman spectroscopy. The CFTS nanoparticles showed photo enhanced super capacitive behaviour which can be used in supercapacitor applications. It occurs due to the excitation of charge carriers with the help of light illumination and significantly helps in increasing the specific capacitance. It is believed that the excess sulphur on the surface plays a crucial role in photo-enhanced defect-activated charge storage in CFTS. The CFTS nanoparticles prepared after the sulfurization process had resulted in a few secondary phases. So, to overcome this problem we fabricated CFTS thin film absorber layer with the help of sol-gel via spin coating technique. Then these CFTS thin films were administered with different types of stabilizers such as lactic acid and monoethanolamine (MEA).

Then the CFTS thin films were subjected to sulfurization process to form required stannite phase as the required phase formation wasn't completed. The sulfurized CFTS thin films were composed of stannite structure and the band gaps in the range of 1.3 to 1.4 eV. CFTS thin films were used for making memory devices known as memristor. Memristor is also known as electrical bistable device. These devices are mainly based on mechanism of ion migration and filament formation. Thin film bistable memory switching devices are those which exhibit two types of different conducting states known as high and low conducting states at particular voltages applied. However, the sulfurized CFTS thin films were devoid of electrical bistability behaviour but showed good photocurrent responses and photostability.

As the sulfurized thin films also resulted in the presence of secondary phase, we adopted a different thin film fabrication method known as SILAR technique to obtain a phase pure CFTS thin films. Then CFTS thin films were synthesised with the help of SILAR method. This fabrication method helped to fabricate a phase pure stannite CFTS thin film, which was then used for the memory applications. Experimental studies suggested that the bistability occurred due to charge confinement of the CFTS grained structure on the thin film. The high-conducting state showed increased capacitance and reduced resistance is calculated from the impedance measurements.

Finally, we also explored the effect of p-n junction in order to use this for solar cell applications. An n-type Bi_2S_3 thin film was fabricated beside the p-type CFTS thin film which forms a p-n junction. We then fabricated polycrystalline $Cu_2Fe_{1-x}Ba_xSnS_4$ (CFBTS) alloy thin films with the help of SILAR method without any post annealing process. The XRD patterns exhibited the structural transition from stannite to trigonal phase with the increasing barium content. The band gaps of the as-prepared CFBTS thin films were found to be increased from ~1.5 to ~1.8 eV, almost in a linear increase trend with the increasing barium content. Finally, a high performance photoelectrochemical (PEC) cells Glass/FTO/CFBTS photocathode with improved stability was fabricated without any addition of bilayers.

Chapter 1
Introduction

CHAPTER 1

Introduction

1.1 Energy crisis: an inevitable aspect

The energy demand across the globe has risen in the recent period. In all other energy sources, the burning of fossil fuels in the form of petroleum and coal accounts for almost 86% of the energy supply [1]. This non-renewable source of energy led to an alarming increase in the emission of harmful carbon dioxide and other greenhouse gases thereby contributing to global warming. The aftermath of this is seen as a drastic climatic change that has contrived differences in human beings and other living organisms. Hence, there is a need for a crucial requirement to develop potential substitutive energy sources that are clean and renewable. Alternative renewable energy sources such as hydroelectric, geothermal, and solar energy meet 10% of the demand. Finding a global solution to these energy problems with little or no damage to the environment is a prime concern of all and anyone in the world. So the need for optimization of these renewable sources of energy has become vital for the human race [2].

The reduction in the use of fossil fuels and the development of a new renewable energy source is the solution to prevent the adverse climatic changes and global distortion of ecological systems. Amongst all other alternative renewable sources of energy, solar energy is considered to be the most extensively used form of renewable energy across the world. As the energy from the sun is not limited, harnessing this energy can lead to a global solution for energy-related problems. The conversion of solar energy can be either by thermal or photovoltaic systems. Our research focus is mainly on two areas. First, on the development of new techniques and materials for the fabrication of low-cost solution-processed absorber layers. Secondly, we draw light on the development of these absorber layers for various multifunctional applications. The fabrication of solution-processed absorber layer avoids the expensive vacuum based synthesis. Also, it helps in reducing the production costs in the manufacturing technology for flexible thin film electronics. Solution-processing allows control over thickness, composition, porosity, doping, substrate coating and band structure engineering. In applications where high crystallinity is not required, then materials may be processed and deposited from solution based techniques instead of using high energy vacuum based methods. Finally the thin films made out of solution-processed methods are even more efficient than those made by vacuum-based methods.

Amongst all the solar cell technologies obtainable to date, silicon solar cells still lead the energy market. The conversion of solar energy can be done by photovoltaic systems in which Becquerel, a French scientist in 1839 discovered the photovoltaic effect on silicon crystals [3]. The first solar cell based on silicon with less than 1% efficiency was first discovered in 1941 by Ohl [4]. Further, much better improvement in the efficiency was reported by Bell laboratory which was the fabrication of the first crystalline silicon solar cell with 4.5% in 1953. Then the discovery of silicon as a semiconducting material further helped in the increase of its efficiency from 6% in 1954 to a record efficiency of 26.3% recently [5, 6]. However, though the availability of silicon is abundant, it is essential to be highly crystalline and pure to attain solar cells of significantly high efficiency. Also, the associated cost with the manufacturing of silicon solar cells is comparatively high as it is a multi-step process.

1.2 Different generation of solar cells

Solar cells are classified into the first, second, and third generation of solar cells. The first-generation solar cells are traditional wafer-based of large-area and capable of generating electrical energy from light sources having a wavelength to that of sunlight. These types of solar cells are made from silicon. They are the highest in terms of efficiency and are currently available for residential use. And it accounts for more than 80 % of the solar cells sold across the world. Monocrystalline, polycrystalline, amorphous silicon cells are the three types of silicon-based cells used for the production of solar panels. It is seen that generally, silicon-based solar cells are more efficient than the other non-silicon based sells. However, the high temperature working capability of this type is a matter of concern as compared to the thin-film solar cells. As the high purity of products are achieved by high temperatures and the volatile impurities are thrown out of the sample.

Thin films are the second generation types of solar cells. They are made from semiconducting materials which are only a few micrometers thick. The much lower cost of manufacturing and the use of fewer materials allow the manufacturers to sell these panels at an affordable low cost. There are three types including amorphous silicon and others are the non-silicon-based materials namely cadmium telluride (CdTe) and copper indium gallium diselenide (CIGS) [7, 8]. These solar cells offer high efficiency but materials like Indium,

Gallium, and Tellurium are rare and expensive, while selenium and cadmium are toxic in nature. And the third generation solar cell includes developing technologies like dye-sensitized solar cells (DSSC) and organic photovoltaics (OPVs). These types of solar cells are potentially able to overcome the Shockley-Queisser limit for single bandgap solar cells [9]. But these solar cells are still in development phase and require extensive work to meet the large-scale manufacturing requirements. Consequently, the research focus on non-toxic, low-cost, and earth-abundant materials is necessary as a good alternative to the previously said types of photovoltaic materials. There are several elements for the advancement of quaternary earth-abundant chalcogenide compounds such as copper zinc tin sulfide (CZTS), copper zinc tin selenide (CZTSe), and CZTSSe. Moreover, CZTS has optical and electronic properties similar to that of CIGS. Also, CZTS is a very promising compound with its efficiencies exceeding 12 % have been already reported [21]. Not even single elements in its constituents are toxic and are five times less expensive than CIGS.

1.3 The idea of low-cost thin films

The motivation to develop thin-film solar cells came from the inception of photovoltaics. For the aim of mass production, the cost of the manufacturing process should be made cheap. The manufacturing of thin films is presumed to be the answer to the low-cost requirement. Materials utilization is an important aspect in the field of photovoltaics as the technologies mature. The thin films can only be inexpensive if the yield is high and the raw materials used are fair to good. Also, it will aid in making thin films achieve higher efficiencies and stability and mainly demonstrating low-cost methods in the actual manufacturing of thin films. Once established in the market, these thin films are likely to make good progress and could even come to dominate the photovoltaic market of the world. The future of thin films will depend on the resources and endurance which are desirable to overcome the challenges. These thin films are fabricated with the help of nano-sized crystalline semiconducting nanoparticles which exhibit excellent optical and electronic properties.

1.4 Semiconductor nanoparticles

Generally, particles within the range of 1 to 100 nanometers (nm) in dimensions are known as nanoparticles. This thesis will discuss more semiconducting nanoparticles. These materials include nano-sized with light-emitting crystalline materials with characteristic properties such as carrier multiplication and spectral diffusion [10]. In a semiconductor, there is an important terminology known as bandgap, which is the energy required to promote an electron from the valence band into the conduction band. The electron is promoted to the conduction band only when a material absorbs the photon with energy greater than the bandgap, leaving behind a positively charged hole in the valence band. An electric field can be applied to move these electrons and holes [11].

Semiconducting nanoparticles have been considered as ideal materials for solar energy harvesting and conversion. These semiconductors are typically fabricated as thin films on the glass substrates having a conducting layer in the solution form. Solution-based synthesis has been optimized to provide control over various aspects such as size and morphology [12]. Also, we can tune the bandgap of these semiconducting nanoparticles with broad absorption and good stability [13].

1.5 Quaternary chalcogenide Cu_2ZnSnS_4 (CZTS)

1.5.1 General properties of CZTS

Among all other earth-abundant photovoltaic absorber materials, the quaternary chalcogenide compound Cu_2ZnSnS_4 (CZTS) has attracted much attention. CZTS has optical and electronic properties that are similar to that of $Cu(InGa)Se_2$ (CIGS). CZTS is mainly used for potential optoelectronic applications such as solar energy harvesting and photoelectrochemical (PEC) cells [14, 15, 16]. The I_2-II-IV-VI_4 semiconductor is a promising material for solar conversion applications because of its direct bandgap of around 1.5 eV, it can absorb most of the solar spectrum [17]. Added significant optical property is the high optical absorption coefficient and high carrier concentrations similar to that of CIGS. CZTS doesn't have any rare element and only consists of earth-abundant materials making it more reliable. This aspect is associated with the scalability and mass production of solar cells as the rare elements hinder the above-mentioned aspects.

1.5.2 Different crystal structures of CZTS

CZTS has three different crystal structures such as kesterite, stannite, and wurtzite structure with space group $I\bar{4}$, $I\bar{4}2m$ and $P6_3mc$ respectively as shown in **Fig. 1.1**. The kesterite phase is derived from the CIGS structure through isomeric substitution of zinc and tin in the place of indium and gallium. In the case of stannite structure, there are distinct layers of zinc cations, but both are in different layers. Whereas, in the case of wurtzite structure the metal cation sites are mixed and the positions of zinc, copper, and tin are randomly arranged. However, kesterite and stannite are similar to zinc blend as the metal cations are located in the tetrahedral bonding positions that are fixed. Among all the three structures, kesterite is the most stable one. The formation energy of kesterite is much lower than that of the stannite and is thereby thermodynamically more stable. This confers that the kesterite CZTS has the calibre of a promising candidate and an important alternative in the application as an absorber layer in the thin-film solar cells compared to those of the expensive CIGS sector.

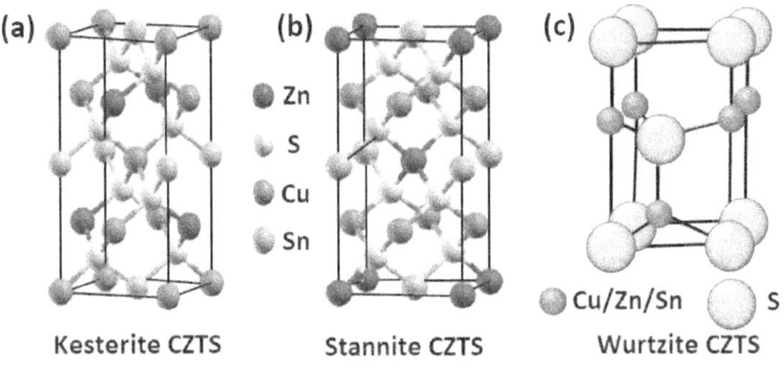

Figure 1.1 Structures of CZTS **(a)** kesterite, **(b)** stannite and **(c)** wurtzite.

CZTS absorber layers can be synthesized with the help of a wide variety of techniques such as sputtering, evaporation, electrodeposition, spray pyrolysis, etc [18, 19, 20]. Solar cells with CZTS have been reported with the best efficiency of 12.6% with hydrazine based solution approach [21]. Along with solar cells, CZTS has shown other valuable applications such as photocatalysis and hydrogen evolution reactions for water splitting applications [22, 23]. One of the fundamental restrictions of the efficiency of CZTS is due to the large V_{oc} deficit and other deformity related factors.

1.5.3 Defect related factors of CZTS

The defect-related factors associated with the growth of CZTS are numerous due to which there occurs a loss of V_{oc}. There are different types of defects such as antisite defects (Cu_{Zn}, Zn_{Cu}, Cu_{Sn}, Sn_{Cu}, Zn_{Sn}, and Sn_{Zn}), vacancy defects (V_{Cu}, V_{Zn}, V_{Sn}, and V_S), and interstitial defects (Cu_i, Zn_i, and Sn_i). These defects happen because of extreme tail states because of electrostatic potential changes brought about by the imperfections in the semiconductors. One of the significant key deformities in the antisite imperfection of CZTS is caused because of CuZn antisite deformity having low formation energy and the proximity of Cu and Zn in the periodic table. The ionic radii of both Cu and Zn are similar and hence it is easy to form these types of defects in CZTS. Thus, it causes Cu_{Zn} and Zn_{Cu} antisite defects that result in severe potential fluctuations and tail states. The presence of Cu and Zn cation issues in the CZTS structure might be the explanation for the poor open-circuit voltage. Moreover, Cu_{Zn} is considered as the lowest energy acceptor and Zn_{Cu} is considered as donor defects in the case of CZTS and CZTSe separately. Accordingly, the antisite pair $[Cu_{Zn}^- + Zn_{Cu}^+]$ with lower formation energy is formed due to their compensation. Subsequently, the arrangement of CuZn antisite defects is effortlessly produced in high concentrations because of this issue. Besides, the presence of point imperfections and deviation from the ideal stoichiometry may bring about self-doping. It can have a critical negative effect on the optical and electrical properties of CZTS. Likewise, we as a whole realize that as the number of elements builds, there is a huge impact on the basic structural property which brings about different intrinsic defects. Various theoretical defect studies based on density functional theory (DFT) calculations have been used to study the defect mechanisms in the crystal structures [24, 25].

1.5.4 Secondary phases in CZTS

It is difficult to synthesis pure CZTS without any secondary phase due to these defects and thus forms binary and ternary secondary phases including ZnS, CuS, SnS, and CuSnS during growth and post-growth processes under different conditions are shown in **Fig. 1.2**. As the stability of the CZTS material is comparatively less and the biggest challenge is to develop a phase pure structure without any secondary phases. The impact of various aspects of secondary phases in CZTS materials such as bandgap, structural, optical, and electrical properties will have a negative impact on solar cell performance. However, these readily formed secondary phases and the inhomogeneity of these secondary phases also contribute to the reduction in the efficiency of the solar cells. Moreover, the Cu-poor and Zn-rich conditions are essential to obtain highly efficient solar cells. Because ZnS secondary phase has the highest probability for the formation of Cu-poor and Zn-rich conditions [26]. CuS is highly conductive and can create a shunting path in the final solar cell fabrication and leads to the lowering of the bandgap energy of the CZTS thin-film solar cells [27]. However, the other secondary phase SnS had no negative effect on the solar cell performance [28]. Accordingly, the replacement of Zn with different metals like Fe, Ba, Cd, Mn could assist with decreasing these deformities. Additionally, the decrease in the antisite defects surrenders by cation replacement has gotten significant consideration.

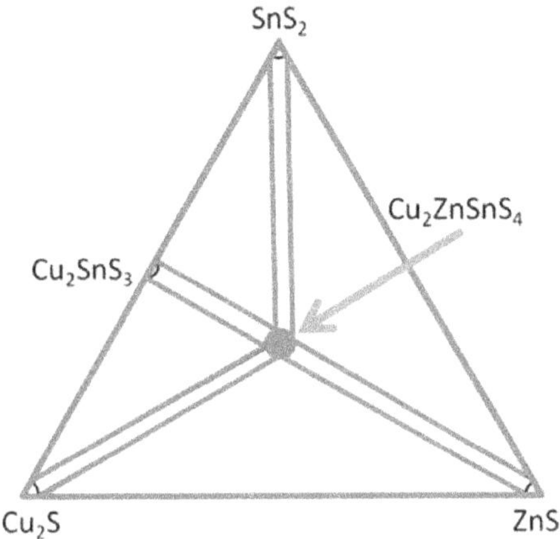

Figure 1.2 Secondary phases associated with CZTS.

1.6 Measures to avoid defects

It is accounted for that 1.4 eV is the ideal bandgap for the absorber materials as the bandgap versus efficiency calculation can be best fitted to the solar spectrum [29]. It is realized that the open-circuit voltage is corresponding to the bandgap of the absorber material. Several studies have indicated that the bandgap tuning from 1.0-1.5 eV can be acquired by changing the sulfur to selenium (S/Se) proportion, which will have the option to improve the productivity of the solar cells. The S/Se proportion controls the bandgap as well as the crystallinity and microstructure of the nanoparticles. Besides, it is hard to control the proportion of S/Se decisively during the tempering cycle known as sulfurization and selenization. Another principle approach in changing the bandgap is the partial substitution of the metal cation and fabricating multi-junctions as possible other options. The creation of a multi-junction will prompt the covering of a whole scope of wavelengths in the solar spectrum. For example, partial substitution of Zn by Cd structural change as well as a bandgap of the absorber material to improve the power conversion efficiency [30].

1.6.1 Partial substitution of cations

Partial substitution is the controlled incorporation of cations into quaternary nanoparticles. Partial substitution of transition metals and alkaline-earth metals into bulk and nanocrystalline semiconductors is a field that has been explored for the past few years [13, 31]. When a system is partially substituted with a new species, the properties of the primary system change drastically, and new and improved applications of the newly synthesized systems are obtained. There are few criteria in which this substitution happens such as the lattice parameters should be reasonably close and the ionic radius of the cation should also be close to the ionic radius of the ion it will substitute. Furthermore, partial substitution is also helpful in the bandgap tuning of the quaternary chalcogenides to a greater extent.

1.7 Cation substitution of Zn

In the nanocrystalline materials, one of the most commonly used types of cations is cadmium, iron, barium, etc. There are many reports in which the partial substitution is studied with the help of transition metal cations as we all know that their unpaired d and f electrons result in interesting and desirable properties. So, to inhibit these problems like severe potential fluctuations, antisite defects, cation disorder, and tail state formations Zn can be substituted with a larger cation. This thesis covers the effect of cation substitution, bandgap engineering, and other photovoltaic applications.

1.7.1 Substitution of Zn with a transition metal – Fe (Cu_2FeSnS_4)

Fe is a metal that belongs to the first transition series and group 8 of the periodic table. It is by mass most the most common element on Earth right in front of oxygen, forming much of Earth's outer and inner core. As a potential replacement, Zn from the quaternary chalcogenide-based semiconductor CZTS was replaced by a transition metal known as iron (Fe) to form Cu_2FeSnS_4 (CFTS). As they are earth-abundant and non-toxic, having an optical bandgap in the range of 1.0-1.5 eV with a higher absorption coefficient ($>10^4$ cm^{-1}) in the visible spectrum range makes them appropriate for using it in photovoltaic applications.

1.7.2 Crystal structure of CFTS

CFTS has two types of crystal structure such as stannite and wurtzite structures and assigned to a space group $I\bar{4}2m$. The structure of CFTS is a tetrahedrally coordiated semiconductor framework where the sulfur anions are bonded with four metal cations. The stannite CFTS structure is like the chalcopyrite and sphalerite structure as appeared in **Fig. 1.3**. There are two Copper, one iron, and one tin ion which are situated at the edges of the tetrahedron [32]. The Mossbauer and magnetic susceptibility studies uncover that the oxidation conditions of Cu, Fe, Sn, and S in CFTS are +1, +2, +4, and -2 respectively [33]. Although, many reports on the possibility of CFTS materials synthesis, only a limited number of studies have been devoted to the fabrication of CFTS based thin-film solar cells.

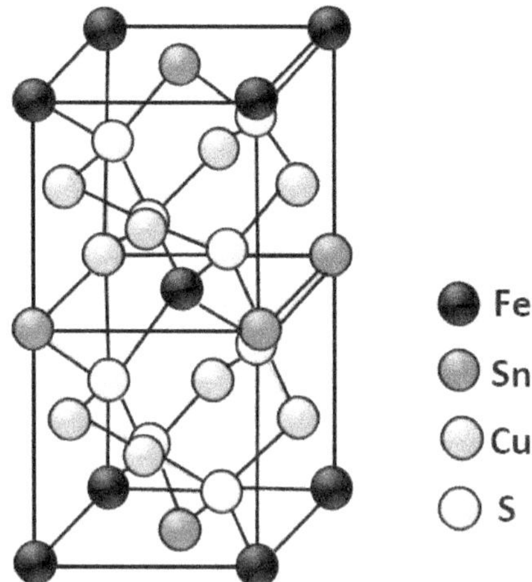

Figure 1.3 Crystal structure of CFTS.

1.7.3 Substitution of Zn with an alkaline-earth metal – Ba (Cu_2BaSnS_4)

Barium (Ba) is a basic earth metal having a place in group 2 of the periodic table of the elements. It is another class of chalcogenide compound semiconductor which has no adverse cation-cation issue. As examined in the past segments, the divergence of imperfections in the kesterite (i.e., CZTS and CZTSSe) solar cells essentially displays further cation antisite defects resulting in non-radiative recombination centers leading to large V_{OC} deficits. As of late, a distant atom mutation concept in CIGSe solar cells was proposed by Z. Xiao et al. This may likewise bring about countless imperfections and deformity groups as the kesterite structure contains three cation elements with the same coordination causing severe band tailing and potential fluctuations. The impact of mutating In by elements like Ba and Sn which are located at different areas of the periodic table are investigated as shown in **Fig. 1.4** [34]. Furthermore, due to the very different electronic properties between Cu and Ba, the formation of the antisite defects will not be conceivable and will bring about upgraded optoelectronic properties.

1.7.4 Crystal structure of CBTS

The earth-abundant orthorhombic $Cu_2BaSn(S, Se)_4$ (CBTS) is recently studied as an absorber material. Ba with an atomic number of 56, and an electronic configuration of $6s^2$, ionic radii of 2.15Å, and atomic radii of 2.53Å. Also, the ionic size of Ba^{2+} (1.49 Å) is much larger than Zn^{2+} (0.88 Å), and replacing these cations in the CZTSSe will not bring about any adverse cation cation disorder.

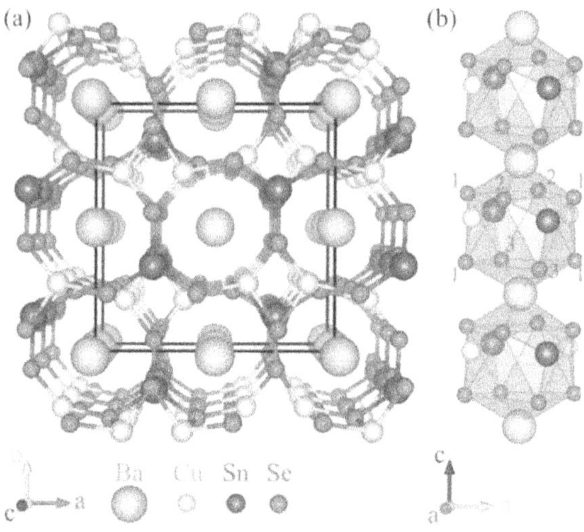

Figure 1.4 (a) Crystal structure of $Cu_2BaSnSe_4$. (b) Local symmetry of the cage-like spaces involved in the tetrahedra channels along [001]. [Reprinted with permission from [34].

The solar cell fabrication using CBTS was at first created by Shin et al. who built up a model solar cell based on trigonal CBTS and reported a PCE of 1.62% [35]. Recently, a surge in the efficiency of CBTS absorber material was observed within a span of few years. Mitzi's group reported a PCE of 5% by decreasing the bandgap near the ideal reach according to the Schokley Queisser (SQ) computation by controlling the Se content [36]. Subsequently, streamlining the Se content just as sulfurization/selenization timing at the same time would help in manufacturing fabricating high-efficiency earth-abundant thin-film solar cells. CBTS has been used as a light absorber for both photovoltaic as well as photoelectrochemical (PEC) applications. CBTS has a large chemical potential window because of large ionic size due to which it is more thermodynamically stable than that of CFTS [37]. For PEC applications, CBTS thin films were fabricated with few overlayers like $CdS/ZnO/TiO_2$ on CBTS thin films as reported by Ge et al. with 7.5 mA cm^{-2} at 0 V vs RHE [38]. Whereas, Zhou et al. recently reported the most efficient and stable photocathode with CBTS/CdS/TiO$_2$/Pt architecture generating a photocurrent of 12.08 mA cm^{-2} [39].

1.8 Literature review

Over the decades, global energy consumption and the human activities that require sustainable energy have prompted research interest in the field of solar energy conversion. With the Sun as the source of energy can satisfy the currently determined need and hence research on the low-cost and highly efficient solar cell materials is viewed as now critical. With this energy source, we can capture, generate, convert, and perhaps store this energy which can be used for addressing mankind's energy challenges in energy conversion and storage [40]. Cu$_2$InGaS$_4$ (CIGS) and CdTe have emerged as the most important chalcogenide semiconductors with an efficiency of 22.6 % and 21.7% respectively [7, 8]. However, the use of expensive and scarce elements such as Indium and Gallium in CIGS and the presence of toxic element Cd cannot be used effectively. Among them, the Cu-based multinary chalcogenides have shown great potentials as next-generation solar cells due to their direct optimum bandgap (Eg=1.2-1.5 eV), high absorption coefficient, and the profusely available inexpensive and non-toxic elements [41].

To supplant these components established researchers have utilized alternative Cu based chalcogenide compounds such as Cu$_2$ZnSnS$_4$ (CZTS) and Cu$_2$ZnSnSe$_4$ (CZTSe) as an absorber material. Furthermore, the power conversion efficiency of this well-known photovoltaic absorber material has reached 12.6% [21]. However, the formation of phase pure CZTS is challenging due to the presence of various defects along with several secondary and ternary phases. These are occurring due to severe tail states due to electrostatic potential fluctuations caused by the defects in the semiconductors. As the ionic radii of Cu and Zn are comparative, the development of these kinds of imperfections in CZTS is a lot simpler. Thus, it causes Cu$_{Zn}$ and Zn$_{Cu}$ antisite defects that result in severe potential fluctuations and tail states. Furthermore, Cu$_{Zn}$ and Zn$_{Cu}$ have the lowest energy acceptor and donor defects in the case of CZTS and CZTSe respectively [42].

Consequently, there is an earnest need to locate the elective components for Zn substitution. So, substituting Zn with a larger cation can inhibit these problems like antisite defects, cation disorder, and tail state formations. As a potential replacement, Zn was replaced by a transition metal known as iron (Fe) to form Cu$_2$FeSnS$_4$ (CFTS). As they are earth-abundant and non-toxic, having an optical bandgap in the range of 1.2-1.5 eV with a higher absorption coefficient ($>10^4$ cm^{-1}) in the visible spectrum range makes them appropriate for using it in photovoltaic applications[43].

Fabrication of superior quality and pure-phase CFTS thin-films has remained a major challenge in the photovoltaic applications to reduce the cost as well. Many vacuum-based and non-vacuum based methods have been tested to prepare CFTS nanomaterials. The former one is considered expensive as it includes vacuum generation for film deposition such as sputtering, pulsed laser deposition (PLD), etc. Whereas the non-vacuum based techniques are comparatively inexpensive as it includes chemical techniques such as successive ionic layer adsorption and reaction (SILAR), spin coating, chemical bath deposition, and spray pyrolysis. Furthermore, in all these studies to fabricate CFTS thin films, major challenges including phase purity, stoichiometric ratio, surface morphology, intrinsic defects, and the uniformity of the thin films are not reported systematically. To overcome these defects, the post-annealing process is necessary. Sulfurization is employed as a post-annealing technique in which the thin films are annealed at a higher temperature in the presence of sulfur powder [44, 45]. Another important aspect is that sulfurization helps in increasing the grain size. As the grain size increases, the carrier recombination decreases, and the diffusion length increases. Hence, sulfurization is employed to grow bigger grains.

CFTS was initially employed as p-type material in a hybrid solar cell by Dong et al. and reported power conversion efficiency (PCE) of 0.29% [46]. Then the optical and electrical properties of CFTS thin films fabricated via RF magnetron sputtering followed by high-temperature annealing in the presence of sulfur known as sulfurization was reported by Meng et al. without any metallic grid [47]. But all of these synthesis methods involved the use of vacuum-based techniques which are considered expensive and complicated. For the fabrication of p-type CFTS absorber material for photovoltaic devices, a solution-based approach was employed. Solution-based technique such as successive ionic layer adsorption and reaction (SILAR) was used by AJ Pal's group to fabricate both n-type and p-type layers. They were able to achieve 2.95% efficiency which was the highest PCE reported for CFTS to date [48]. CFTS can be used for other applications as well such as a counter electrode for a dye-sensitized solar cell (DSSC) application as a potential replacement of the expensive platinum as a counter electrode and reported significant efficiency of 8.03% [49, 50].

1.9 Multifunctional applications of CFTS

With the invention of nanoscience and nanotechnology, researchers are developing new functional materials combining all the specific properties. Nanotechnology deals with various fields and finds active applications in energy conversion and storage-related applications such as solar cells, photodetectors, new materials based on electrical bistable devices, electronics, supercapacitors, etc [51, 52, 53]. The development of new materials for energy conversion and storage has broadened their utilization into areas where their semiconducting and conducting properties have encouraged use in many novel applications are shown in **Fig. 1.5**.

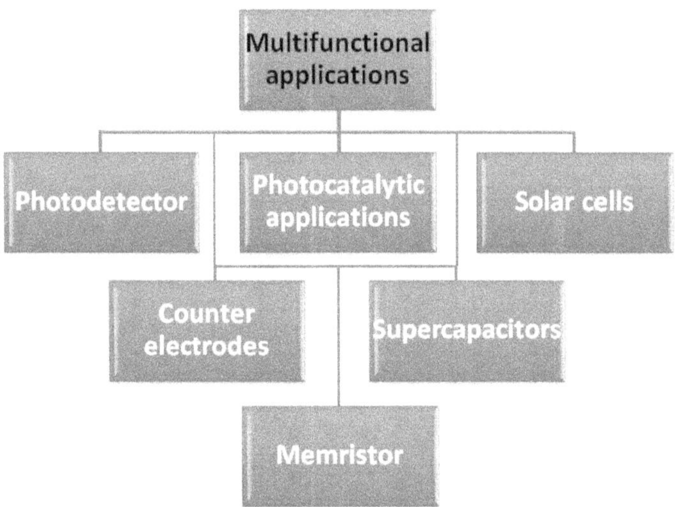

Figure 1.5 Classification of the multifunctional application of CFTS.

CFTS also points in the direction of some application-level aspects such as it is widely in Dye-sensitised solar cell (DSSC) as a used as a counter electrode, photoelectrochemical cells (PEC), solar cell device fabrication, electrical bistability for memory devices, etc. In the case of DSSC, device fabrication involves several steps and some key components as well. These components include a photoanode, sensitizer dye, redox mediator, and counter electrode. In some cases, the counter electrodes being used are of high costs, like the platinum-coated glass. To reduce the cost without compromising with the efficiency, CFTS was used as an alternate material.

The quaternary chalcogenide-based counter electrodes are emerging as one of the better replacements of the platinum-based materials. The low fabrication cost and facile preparation at relatively low temperatures have been used for the development of low-cost photovoltaics. Dye-sensitized solar cells (DSSC) are promising devices with reasonable high power conversion efficiency. DSSC consists of a mesoporous TiO_2 which is loaded with a dye known as the photoanode, an iodide/triiodide redox couple as an electrolyte, and a counter electrode. The counter electrode (CE) collects electrons from the external circuit to reduce I_3^- to I^-. Commonly, platinum is used as a CE, due to its high electrocatalytic applications. However, Pt poses a significant drawback owing to the high cost and scarcity, and hence we are in dire need for a potential replacement with alternative low-cost CE materials. Therefore, there is a strong need for the advancement of earth-abundant materials that helps in reducing the cost. Recently, Jae-Young et al. in 2013 demonstrated the use of different chalcogenide compounds such as CuS, $CuSnS_3$, CFTS, and CZTS as counter electrodes. Among all these CEs, the device fabricated with CFTS as CE was the best performing device with 5.6 % as the power conversion efficiency.

Another solution-processed method followed by sulfurization was employed by K. Mokurala et al to fabricate CFTS thin films for DSSC. With the help of the solvothermal process, the CFTS nanoparticles were synthesized which were dispersed in toluene to get a homogeneous slurry. This was then coated over an FTO substrate to fabricate a CE followed by sulfurization at 550 °C. Then DSSC was assembled to exhibit a power conversion efficiency of 7.1 %, but lower than the Pt-based counter electrode (8.2 %) [54]. In this study, K. Mokurala et al. reported the replacement of Pt with CFTS to obtain higher efficiency, but it could report only lower value when compared to Pt.

However, there are other studies reported, which proved that CFTS is a better replacement for Pt. That is in 2014, RR Prabhakar et al. synthesized CFTS thin film by spray pyrolysis method followed by sulfurization at 400 and 500 °C labeled as CFTS4 and CFTS5. These, when used as a CE, exhibited a record power conversion efficiency of 6.78 and 8.03 % for CFTS4 and CFTS5 respectively when compared to Pt (7.57 %) and its JV curve is shown in **Fig. 1.6** [50]. The presence of the second phase causes poor catalytic activities and lowers the efficiency of CFTS4. This being the highest reported value and suggests that CFTS can be a feasible promising candidate for the substitution of Pt as CE in DSSCs.

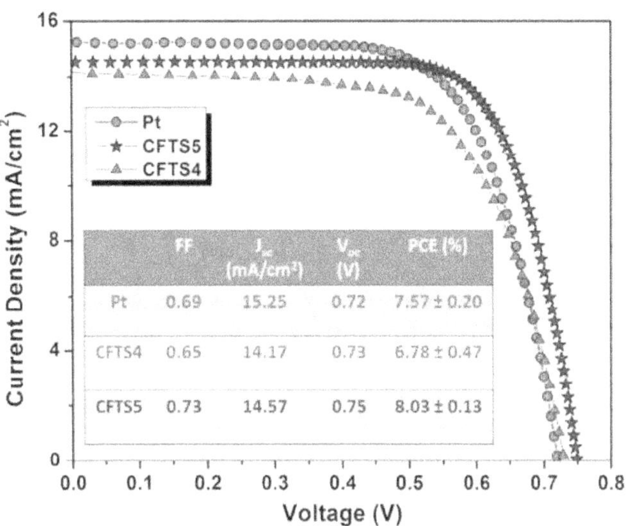

	FF	J$_{sc}$ (mA/cm²)	V$_{oc}$ (V)	PCE (%)
Pt	0.69	15.25	0.72	7.57 ± 0.20
CFTS4	0.65	14.17	0.73	6.78 ± 0.47
CFTS5	0.73	14.57	0.75	8.03 ± 0.13

Figure 1.6 Current density (J) – Voltage (V) plots of DSSCs fabricated using Pt and CFTS4 and CFTS5. Reprinted with permission from [50].

These earth-abundant materials ever since their time of introduction has reduced the cost as well as it have reduced the time for applying it in practical applications. The solution-processed CFTS thin films were deposited with some of the non-toxic earth-abundant precursors with the help of the spray pyrolysis technique. This generated power conversion efficiency as high as 8.03%, which was relatively better efficient than the platinum-based counter electrodes. The different sulphurization temperatures have also experimented and the desired electrical and optical properties of these solution-processed CFTS materials indicate that these thin films can be used in solar energy harvesting devices as an absorber layer [50].

The CFTS materials synthesized via solvothermal method can also be utilized for replacing the platinum counter electrodes. That is, the quaternary chalcopyrite sulfide semiconductors Cu_2XSnS_4, where X can be replaced by earth-abundant low-cost materials like Fe, Co, Ni, Zn, etc and the steps involved in the preparation is shown by the schematic diagram as represented in **Fig. 1.7**. The solution-processed quaternary chalcopyrite sulfide exhibited high catalytic activity for the electrolyte reduction in the ZnO based DSSC. As the bandgap of these earth-abundant materials falls in the range of 1.0 to 1.5 eV, it can be used for solar energy harvesting applications also can be used to replace platinum-based electrodes in low-cost DSSCs [55].

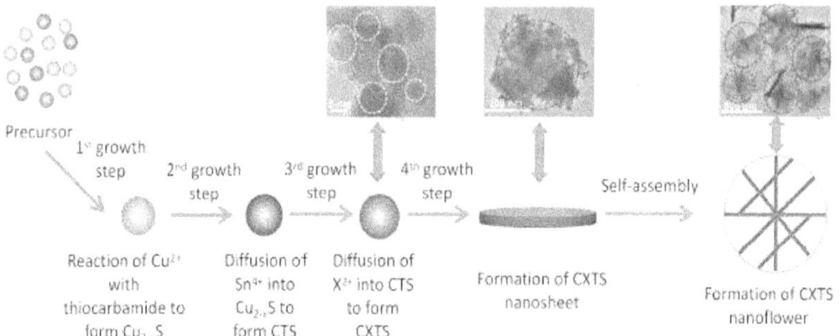

Figure 1.7 Schematic formation of Cu2XSnS4 (X=Mn, Fe, Co, Ni, Zn, Cd) nanoflower. Reprinted with permission from [55].

Photocatalysis is a term that defines the change in the chemical transformation rate under the action of light in the presence of a catalyst that absorbs light. Various types of absorber layers with suitable band gap and absorption coefficients have been attracted recently. Among them, the most promising material is the chalcogenide material CFTS, which is composed of non-toxic and abundant materials. The very recent studies include CFTS prepared by spray pyrolysis technique using different thiourea concentrations such as 4, 5, 6, and 7.10^{-2} M by C Nefzi et al. Among these concentrations, 6.10^{-2} M exhibited 76 % photodegradation of methylene blue (MB) dye at 4 hours which was considered highest degradation [56]. Then the same group in 2020 fabricated CFTS by the same method and studied the photocatalysis application. They reported that CFTS thin films could degrade 81% of MB dye in just the same amount of time reported in their previous work. This was achieved by changing the substrate temperature from 160 to 280 °C. From those conditions, 280 °C fabricated CFTS thin film exhibited 1.46 eV bandgap energy with a higher absorption coefficient [57].

Another dye known as rhodamine B dye was also tested for photocatalytic activity, in which CFTS powder was used, which was prepared by simple and low-cost ball milling procedure. Here also the bandgap matched for the photovoltaic applications i.e, 1.42 eV. And it was found that the dye got significantly degraded within 120 minutes under the light illumination [58]. Another dye was methyl red (MR) degradation. It is an important monoazo dye which is mainly used in the textile industries. Here, CFTS nanostructures such as nanoflakes and nanoparticles are fabricated by a hydrothermal route at different durations. Among them, nanoflakes were better as it could degrade about 74 % of the MR dye in just 3 hours under light illumination. As the nanoflakes were found to be having smaller grains of varying size that the nanoparticles which are prepared by keeping for a prolonged duration. The nanoflakes were quite stable and can be used again and again.

Another main aspect of any material is its execution in the day to day livelihood where it can be practically applied for human benefit. So, device fabrication is the way in which this feat can be achieved. There are several ways in device fabrication, among which a two-step synthesis discussed earlier (RF magnetron sputtering succeeded by sulfurization) and SILAR are the best suitable technique. Meng et al. investigated the optical and electrical properties of CFTS along with a glass/Mo/CFTS/CdS/i-ZnO/AZO-structured device. Also, no other metal grids were deposited on the device, thereby enabling it to display a power conversion efficiency of 0.07 %. To improve the efficiency of this type of solar cell, the absorber layers, and the physical and annealing parameters must be optimized [47].

In 2016, the first device based on the SILAR method came into existence where both the p-type and n-type layers were deposited with the help of SILAR. The adoption of the low-temperature process and also vacuum-free process has made it economical. The films made through this process were found to be in a single phase with phase purity and is used for large area deposition as well.

Another device was fabricated with the help of another solution-based method known as spin-coating the CFTS precursor solution followed by low-temperature thermal annealing. Here they reported a bandgap of 1.53 eV and high absorption coefficient. Here they used an n-type CdS layer with a total device thickness of 125 nm. C. Dong et al. were able to produce only 0.08 % efficiency with their solution-processed and much simpler structure. The thickness of the absorber layer was also varied and found that, as the CFTS thickness increases the current density decreases linearly [59]. The presence of a secondary phase would lead to deterioration of the power conversion efficiency of a device. The n-type semiconductors used in this study were CdS, Bi_2S_3, and Ag_2S with a bandgap of 2.4 eV, 1.3 eV, and 1.1 eV respectively. Based on this bandgap and band alignment, the schematic of the energy level diagram of these pn-junctions is shown in **Fig. 1.8.**

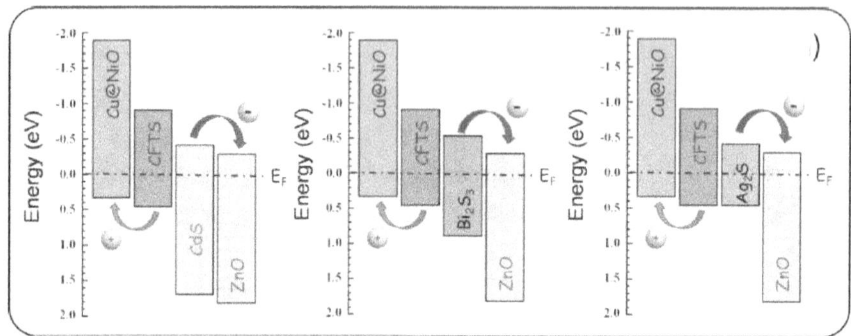

Figure 1.8 Schematic energy level diagram of CFTS|CdS, CFTS|Bi$_2$S$_3$, and CFTS|Ag$_2$S heterojunctions. The line at 0 V represents the Fermi energy after contact. Reprinted with permission from [48].

These devices could generate a power conversion efficiency of 1.37 %, 2.95 % (**Fig. 1.9**), and 0.77 % respectively as well with a high reproducibility made through the maiden approach of preparing a single-phase and highly crystalline CFTS thin-films [48].

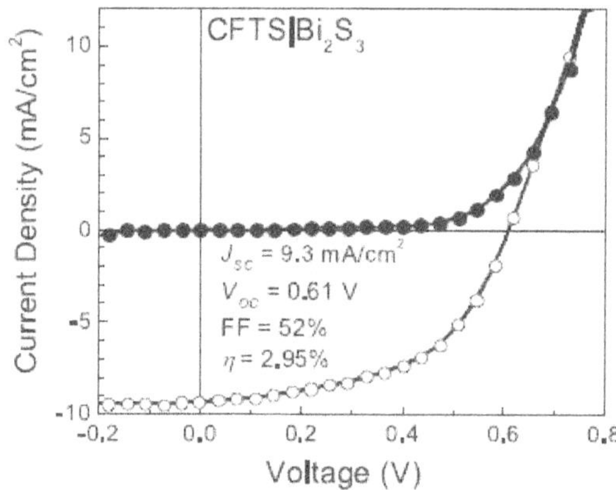

Figure 1.9 *I-V* characteristics of a typical (a) CFTS|Bi$_2$S$_3$, all under dark and 1 Sun illumination conditions. Reprinted with permission from [48].

There are many n-type materials available for completing a pn-junction for fabricating a photovoltaic device. Not only Bi$_2$S$_3$ which will act as an n-type layer but also other n-type materials like ZnS will be used for developing a pn-junction. Using ZnS coated on ZnO nanorods the electrochemical properties, as well as photovoltaic performances of the CFTS thin-films solar cells, have been studied by Anima et al. for the first time and the device structure. And the gold electrodes were used for the top electrode contact and investigated under AM 1.5 standard illumination conditions and displayed a power conversion efficiency of 2.73%. The whole procedure for the synthesis of this device is seen as a schematic diagram as shown in **Fig. 1.10**. The presence of ZnS on ZnO has led to the increased visible light absorption and the enhancement of light has occurred due to the ZnO nanorods alignment. Also proved to be exhibit superior electrochemical properties that may be used for other energy harvesting and storage applications [60].

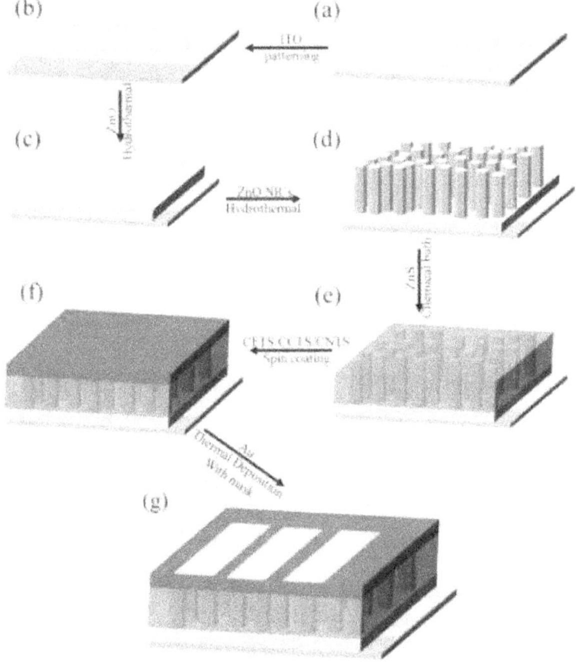

Figure 1.10 Schematic diagram of device structure: **(a)** ITO coated glass substrate, **(b)** ITO patterning **(c)** ZnO seed layers **(d)** ZnO Narorod arrays, **(e)** ZnS coated ZnO nanorod, **(f)** deposition of absorber layer (CFTS/CCTS/CNTS) on ZnS coated ZnO nanorod, **(g)** complete proposed device structure with top layer Au. Reprinted with permission from [60].

In other methods of fabrication involves the process of sulfurization and selenization. Also, the power conversion efficiency can be enhanced with the help of these techniques. Moreover, doping with sulfur or selenium has resulted in the enhancement of efficiency, which is limited to a certain particular area. The high-temperature treatment under inert gas reduced the loss of Sn and S from these films and there is a significant enhancement in the power conversion efficiency of the solar cell due to the Se doping [61]. The synthesis of CFTS nanocrystals by Chao et al. was discussed earlier in the synthesis section, they also fabricated a device structure that delivered a power conversion efficiency of 0.29 % and they used the synthesized CFTS nanocrystals in hybrid polymer-based solar cells (HPSCs) as an electron acceptor material as well. The *I-V* curve along with the device architecture and the band level alignment of this type is shown in **Fig. 1.11** [46].

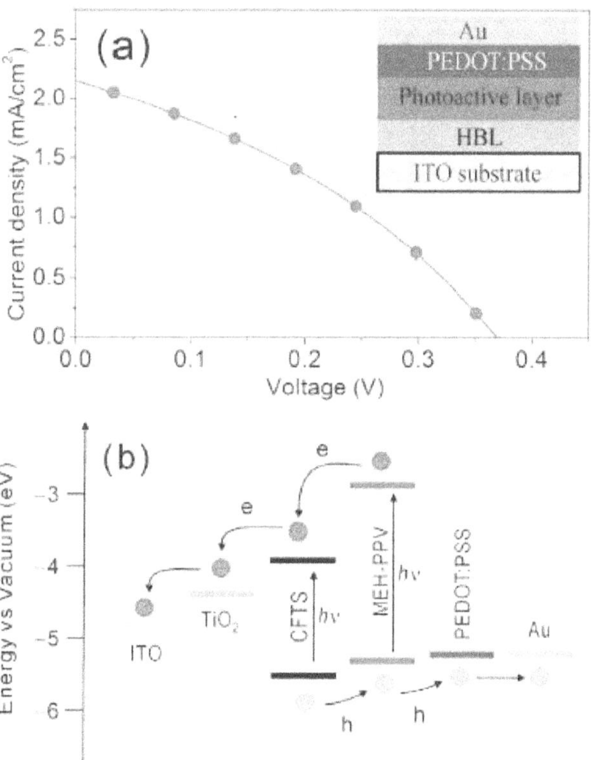

Figure 1.11 (a) J–V curve and **(b)** band level alignments of MEH-PPV/CFTS solar cells. The inset to (a) is device architecture, where HBL means the hole blocking layer. Reprinted with permission from [46].

1.10   Energy and memory-related applications

Similar to energy conversion, energy storage is also an essential element in modern technology. Supercapacitors are considered promising energy storage devices due to their high power density, low cost, and high stability. Quaternary metal sulfides CFTS are an interesting class of materials that can be used for multifunctional applications.

A memristor is a two-terminal device whose resistance depends on one or more internal state variables of the devices. Using memristors one can achieve circuit functionalities, which are not possible to establish with resistors, capacitors, and inductors. The potential applications of memristors are as follows:

- Spintronic devices
- Ultra-dense information storage systems
- Neuromorphic circuits
- Programmable circuits

Memristor was first theoretically established by L.Chau in the 1970s [62]. And later realized that it possesses a unique memory profile by Williams and co-workers [63]. A memristor is also known as an electrical bistable device. These devices are mainly based on the mechanism of ion migration and filament formation. Thin-film bistable memory switching devices are those which exhibit two types of different conducting states at a particular voltage. Depending on both direction and magnitude of applied voltage, the device can be switched from a low conducting state, which is known as the "0" state or "Off" state to a high conducting state, that is, "1" or "On" state. This process may be known as the writing process. This high conducting state can remain stable until a bias voltage is applied and can be reversed back to previous state at a lower reading voltage [64]. The reverse process is achieved by applying a reverse voltage when the conductive state switches back from a high conducting state to a low conducting state. And this process is known as the erase process.

These devices which are functionalized with erasing and writing process can be used in the random access memory (RAM) and flash memories [65]. Some switching materials show write-once-read-many times (WROM) switching property and those can be used as read-only memory (ROM) devices.

For the commercial use of data storage, the device should satisfy the following requirements:

- The small dimension of the memory cell, so that it can be incorporated into integrated circuits.
- Room temperature operation, Low activation voltage to save energy.
- Long retention time and durability.
- High On/Off ratio, short response time.

The multifunctional properties of CFTS have led to employ CFTS as a memory device. With the existence of high and low resistance in a voltage sweep, then there occurs the electrical bistability or a memristor. The co-existence of negative differential resistance (NDR) and resistive switching (RS) is a promising property of a material that can be used as memory devices. CFTS thin films can act as an electrical bistable device which is also explored in this thesis. This work provides an insight into the memory applications of CFTS nanostructures as well.

1.11 Effect of sulfurization

Sulfurization is the process whereby a surface of a material is exposed to the sulfur-containing atmosphere at high temperature. The surface of the film reacts and forms a new compound incorporating sulfur. It can be configured for sulfurization processing using either a resistance heated furnace or an infra-red lamp furnace for rapid thermal processing. Sulfurization processes are used in the production of CFTS thin-film solar cell modules. They are also used for producing 2D layers of transition metal dichalcogenide (TMD) materials for advanced research. The schematic of the sulfurization process is shown in **Fig. 1.12.**

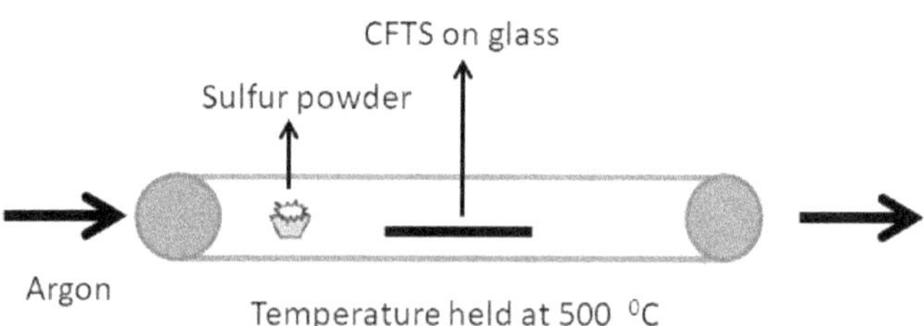

Figure 1.12 Schematic of sulfurization process.

As of late, Xiaohui et al. prepared quaternary CFTS thin-film by electrochemical deposition method followed by sulfurization at two different temperatures under the nitrogen atmosphere. The different characterization studies proved its stannite structure and also its crystallinity improved with an increase in sulfurization temperature. Finally, after the post-sulfurization treatment at 500 °C and 550 °C, the direct bandgap of both the CFTS thin-films was found to be 1.35 eV and 1.40 eV [66].

The formation of impurity-free pure phases in CFTS is observed after a high-temperature annealing process known as sulfurization. It can be performed with the help of a rapid thermal process (RTP) which can be done with the help of an automated control system that can supply a few hundred degrees per minute. Thereby helps in the reduction of the processing time and fabrication procedures. X. Meng. et al in 2015, fabricated pure phase CFTS with tetragonal structure with 0.97 tetragonal distortion (c/2a) [47]. This type of calculation depends on the quality of the CFTS thin films. Whereas, there occurs a change of the structure from rhodostannite to stannite with the increasing sulfurization temperature as reported by the same team. The Sn content is also decreased with the increase in sulfurization without any impure phases [44]. A stannite CFTS structure and grain growth were observed for the sulfurized samples which are heated at a slow rate. Also, the enhancement in the crystallinity is reported after the decrease in the heating rate by X. Meng et al. [45]. Not only the duration of sulfurization time, has that depended on the formation of the pure phase but also the sulfurization temperature. Prabhakar et al. reported that the increase in the sulfurization temperature decreased the intensity of $Cu_2Sn_3S_7$ (CTS) peaks as shown in **Fig. 1.13** and also increase the peak intensity of CFTS thin films as reported by X. Miao et al. [50, 66].

There have been an umpteen number of reports on the effect of temperature and time of sulfurization. However, there were no reports until 2017 on the electrical and electrochemical properties of CFTS. It resulted in the desired stoichiometric ratio due to the evaporation of Sn and S from the films at higher temperatures. The electrical properties such as the resistivities were smaller than that of the as-prepared and the mobility of sulfurized films were higher [67]. The crystallinity of the CFTS thin films is also a matter of consideration for its better performance, thus sulfurization helps in improved crystallinity of the layers and made the surface homogeneous, rough, and dense with many micro aggregates.

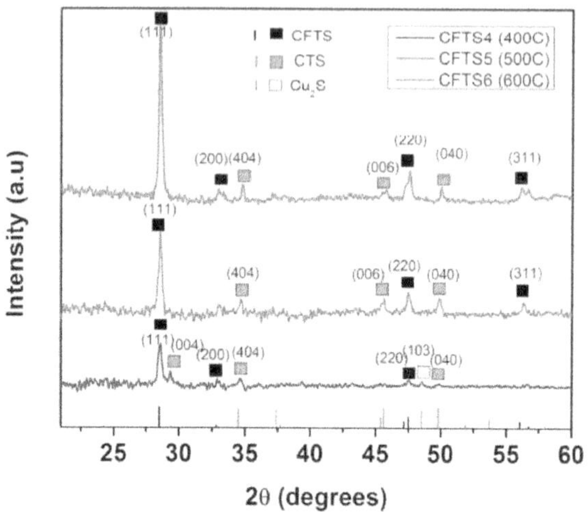

Figure 1.13 XRD patterns of CFTS4, CFTS5, and CFTS6 indicating the presence of the CFTS phase. Reprinted with permission from [50].

H. Oueslati et al. fabricated CFTS by thermal evaporation method on Mo substrate, which was followed by sulfurization. It was noticed that CFTS thin films were converted into the pure stannite crystal structure as shown in **Fig. 1.14 (a)**. Also, it resulted in a compact, homogeneous, and dense CFTS surface after sulfurization as shown in **Fig. 1.14 (b)**. The Raman spectra also matched well with a symmetry mode of the stannite CFTS phase. This deposition technique followed by sulfurization was used for Schottky diode fabrication for the first time with a structure of Al/p-CFTS/Mo [68].

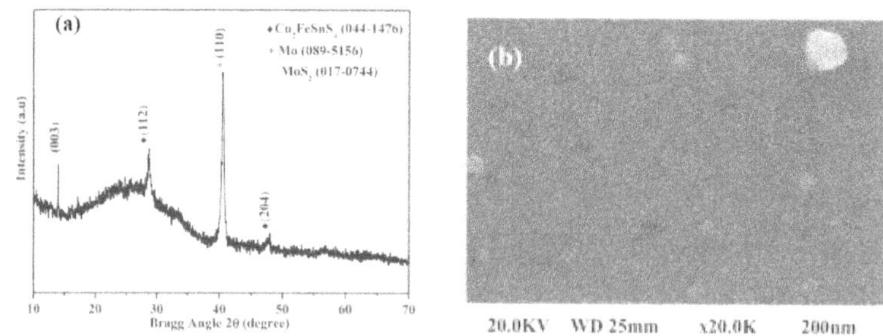

Figure 1.14 (a) XRD pattern of Cu_2FeSnS_4/Mo thin film and **(b)** SEM micrograph of Cu2FeSnS4/Mo thin film. Reprinted with permission from [68].

There are other ways to do sulfurization other than using sulfur powder, which is by treating it with H_2S or S vapor without any need for sophisticated equipment. However, the CFTS thin-films were deposited with the help of a spray pyrolysis technique followed by annealing at different temperatures. From this study conducted by Adelifard et al, it was understood that the electrical conductivity and the hole mobility of this CFTS thin-film increased with an increase in the substrate temperature. And the morphology of the film became uniform and larger gain sizes were obtained with increased temperature. The bandgap energy for this CFTS thin-film increased from 1.42 - 1.55 eV which makes them promising candidates for photovoltaic applications [69].

1.12 Fabrication of CFTS

There are different deposition methods either physical which employs a vacuum or chemical which does not make use of vacuum. Chemical-based techniques are not expensive and are suited techniques for the manufacturing of a photovoltaic system and low cost potential. Chemical techniques are simple, economic and convenient for the deposition of a variety of materials, for example, metal chalcogenides. The main goal of the use of different deposition method is for finding an economic and suitable route of deposition. In this work we made an attempt to employ solution-processed non-vacuum based techniques such as successive ionic layer adsorption and reaction (SILAR), hydrothermal, sol-gel, spray pyrolysis, chemical bath deposition (CBD), etc.

1.12.1 CFTS fabrication by non-vacuum based techniques

Here we discuss the recent development in CFTS and various synthesis methods to prepare CFTS nanoparticles. The different synthesis methods were approached by researchers all over the world and it includes successive ionic layer absorption and reaction (SILAR), solvothermal, sol-gel, sprays pyrolysis methods, electrospinning, etc. CFTS nanoparticles having different morphologies like microspheres, nanospheres, nanorods, nanoflowers, etc synthesized via solvothermal have been reported several times using various solvents such as ethylene glycol (EG), N,N-dimethyl formamide (DMF), Dimethyl sulfoxide (DMSO), etc [70, 71, 72].

Then in the case of other simple chemical methods which involve SILAR technique, one of the promising way of film-fabrication procedure that offers excellent film quality and also low-cost method. Thin films made out of this form also reports single-phase and highly crystalline CFTS thin-films [48]. The SILAR method was first reported by Ristov et al. in 1985 [73]. SILAR method is mainly useful for the deposition of thin films of chalcogenide groups such as binary, ternary and quaternary chalcogenides. This technique is essentially based on the adsorption and reaction of the ions from the cationic and anionic solutions. After each immersion process, a rinsing process with distilled water (DI) is followed to prevent the formation homogeneous precipitation in the solution. The thin films are obtained by the adsorbed cations on the substrate followed by reacting with appropriate anion precursor solutions.

The hydrothermal method is used for the synthesis of various nanoparticles that depend upon the solubility of minerals in hot water and under high pressure [74, 75]. The growth of this nanoparticle is performed in an apparatus consisting of a high pressure steel vessel called an autoclave, in which the precursor is supplied along with water. If the solvent is an aqueous solution, then this method is called hydrothermal, otherwise, it is known as solvothermal.

The synthesis of stannite and kesterite phases has been used recently in applications like solar cells, due to the possibility of employing it in the field of photovoltaics. The phase transition of stannite and kesterite phases have been started investigating since 1972 and is of great interest in the modern period as well. And the incorporation of the transition elements like copper, iron, and zinc in the stannite and kesterite series were also studied by Bonazzi et al. in 2003 [76]. Schorr et al. investigated the cation restructuring in the stannite and kesterite mixture in 2007 with an idea to study the distribution of Cu, Zn, and Fe at various positions [77]. However, the synthesis of Cu_2FeSnS_4 nanocrystals through thermal reactions of metal precursors along with the optical properties have been studied in 2012 by Liang et al. Bandgap associated with synthesis was 1.33 eV, which is considered as a possible material for solar cell applications [78]. The zincblende CFTS structures were synthesized with the help of solvothermal reaction, where the synthesized CFTS particles displayed a spherical morphology and it also exhibited ferromagnetic behavior. The bandgap associated with this process was also in the range of 1.2 to 1.5 eV which is suitable for photovoltaic applications [79].

As we mentioned earlier, the synthesis of CFTS involves various methods, and one of the common techniques is the solvothermal method where both the zinc blende and wurtzite structures were synthesized. Zhang et al. investigated the synthesis of CFTS nanocrystals through a solution-based method with a tunable crystal phase in 2012. Also, the wurtzite CFTS structure synthesized in this study reveals the improved stoichiometry configuration and makes it possible to tune the Fermi energy level to wider ranges because of its unsymmetrical cation distribution. The phase-controlled synthesis of CFTS through solution-based methods resulted in spheroids structure with d-spacing matching the wurtzite structure. The morphology of the wurtzite -phase CFTS nanocrystals showed some oblate spheroids, whereas zinc blende CFTS nanocrystals exhibit a triangular nanoplate. The bang gap of wurtzite and zinc-blende is 1.46 eV and 1.54 eV respectively which is suitable for low-cost thin-film solar cells. Among this zinc blende phase is more stable and is favored at high temperatures with particular morphology for both the structures [80].

With the aid of the solvothermal method, Chao et al. synthesized CFTS nanocrystals of 3-7 nm size range in the presence of amine and other solvents. In this study, he used these synthesized CFTS nanocrystals were used in hybrid polymer-based solar cells (HPSCs) as an electron acceptor material for which the device fabrication is explained in the upcoming segments [46]. In a particular synthesis based on solvothermal reaction stannite and wurtzite, CFTS thin-films were synthesized directly onto the glass substrate. Also, the other significant electrical properties like carrier concentration and carrier mobility were studied. The band gaps of stannite and wurtzite structures were 1.3 and 1.34 eV respectively which is ideal for thin-film solar cells and also concluded that the two different compositions of solvent play a key role in the formation of two different phases of CFTS [81]. Also, the synthesis duration plays an impact on the photocatalytic properties of CFTS nanoparticles.

1.12.2 CFTS fabrication by vacuum-based techniques

In another case, a single one-step thermal decomposition of metal precursors was adopted for the first time to synthesis CFTS nanoparticles in 2014. The bandgap associated with these CFTS nanoparticles was optimal for use in photovoltaic applications (1.40 eV). This method is also considered as a low-temperature budget method for the preparation of thin-films with a single-phase [82]. Not only the single-step process is used but also, a two-step method is employed to synthesis CFTS thin-films. RF magnetron sputtering of precursors was deposited sequentially from tin, iron and copper targets respectively in the presence of high purity Ar gas atmosphere (20 sccm). The sputtering pressures were maintained at 1.2, 1.6 and 1.6 Pa for Sn, Fe and Cu layers. It is then followed by a post-annealing process with a particular optimized sputtering pressure for each precursor followed by a sulfurization at 550 °C. The bandgap of this CFTS was matching with the ideal bandgap according to the Shockley-Queisser limit (1.42 eV) and the obtained CFTS was tetragonal structure [47]. The microwave irradiation method conducted by Guan et al. used to synthesis CFTS particles has resulted in a flower-like appearance as shown in **Fig. 1.15**. The bandgap associated with the CFTS particle was 1.52 eV, which was in close range for utilizing it for the absorber layer applications in photovoltaics [83].

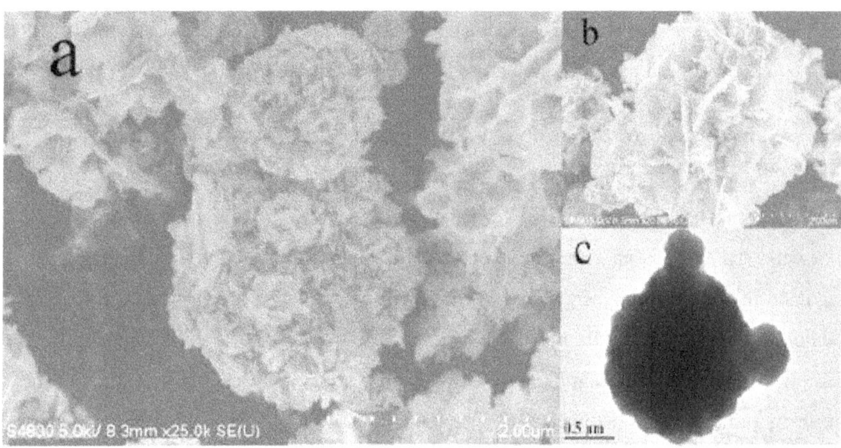

Figure 1.15 (a) Low, **(b)** high magnification SEM images of CFTS particles and **(c)** TEM images of CFTS particles. Reprinted with permission from [83].

One of the other chemical methods is the aerosol assisted chemical vapor deposition (AACVD) of the precursors for the synthesis of CFTS. This method is also used for the deposition of other metallic salts including Ag, ZnO, SnS, Fe_2O_3, etc. In this article, Kevin et al. carried out the thermal decomposition studies with a particular heating rate in the presence of nitrogen. TGA measurements and other electrical and optical properties were also studied in this work. Based on the stoichiometry ratio and other properties, at 350 °C the best films were obtained and the irregular plate-like structure is revealed by TEM analysis followed by elemental mapping showed the uniform composition of elements. The bandgap value was in the range of 1.1-1.6 eV, which was suitable for solar cell materials [84].

Recently, a vacuum thermal evaporation technique was used to coat CFTS onto a glass substrate followed by sulfurization at 400 °C. In one of the study CFTS was used for making a Schottky diode and the other was used to partially substitute Zn to form $Cu_2Zn_xFe_{1-x}SnS_4$ (CZFTS) [68, 85]. The structural analysis confirmed the pure stannite phase and an Al top electrode resulted in the formation of the junction to form a Schottky diode. And the rectification of the curve as shown in IV characteristics under the forward and reverse bias confirms that the junction behaves like a Schottky diode with Al/p-Cu_2FeSnS_4/Mo as shown in **Fig. 1.16 (b)** and **(a)** respectively. Finally, it resulted in a saturation current of $4.89.\ 10^{-5}$ A with an ideality factor of 1.48. Whereas, in the partial substitution report, the CFTS formed also matches well with the stannite phase without any secondary phases after sulfurization [68].

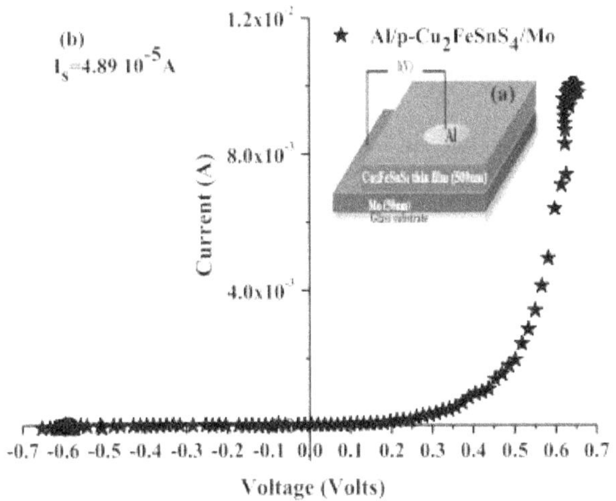

Figure 1.16 (a) The device structure and **(b)** The experimental current-voltage characteristics of the Al/p-CFTS/Mo Schottky junction. Reprinted with permission from [68].

1.13 Deposition techniques

There are different deposition methods either physical which employs a vacuum or chemical which does not make use of vacuum. Chemical-based techniques are not expensive and are suited techniques for the manufacture of a photovoltaic system with low-cost manufacturing. For the deposition of a variety of materials such as metal chalcogenides, chemical methods are simple, economic and convenient. The main goal of the use of different deposition method is for finding an economic and suitable route of deposition. Here we are focussing on solution-processed non-vacuum based techniques such as successive ionic layer adsorption and reaction (SILAR), hydrothermal, sol-gel, spray pyrolysis, chemical bath deposition (CBD), etc. The deposition techniques that are used in this thesis for CFTS fabrication are represented as a flow chart in **Fig. 1.17**.

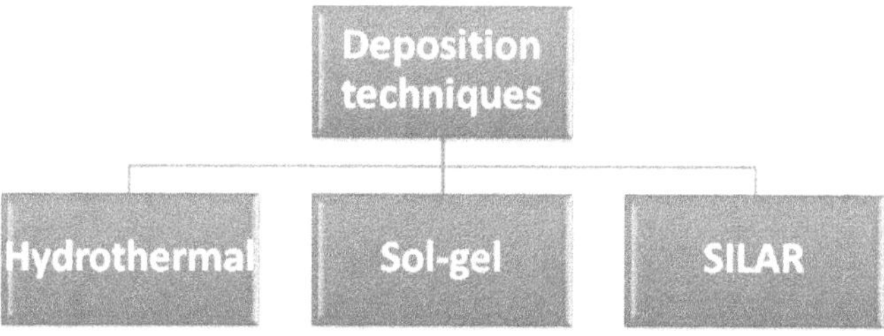

Figure 1.17 Deposition techniques used for CFTS fabrication.

1.13.1 Successive ionic layer and adsorption reaction (SILAR)

In recent years, metals, semiconductors, conducting polymers, etc. in the form of thin films have been synthesized by the SILAR method. Among all the thin film deposition methods, the SILAR method is relatively simple and offers a wide scope of advantages over other expensive methods of thin film deposition. It is considered as one of the newest solution methods for the deposition of thin film, it is also known as a modified version of chemical bath deposition (CBD). In this process, the deposition rate and thickness of the film can be controlled by changing the deposition cycles. From this technique, relatively uniform films can be attained on any substrates having any shape and size. The main advantage of this method is that it does not require any expensive and sophisticated equipment at any stage of the fabrication process. Furthermore, this method requires lower operating temperatures and can prevent interdiffusion and dopant redistribution.

The SILAR method was first reported by Ristov et al. in 1985 [73]. SILAR method is mainly useful for the deposition of thin films of chalcogenide groups such as binary, ternary and quaternary chalcogenides. This strategy is for the most part dependent on the adsorption and reaction ions of the particles from the cationic and anionic solutions. After every immersion cycle, a rinsing process with distilled water (DI) is followed to prevent the formation homogeneous precipitation in the solution. The thin films are obtained by reacting with appropriate anion precursor solutions. Adsorptions of the arrangements are normal when both the anionic and cationic solutions having heterogeneous phases are gotten contact with one another. Adsorption is considered as an exothermic cycle and goes about as a surface phenomenon among ions and the surface of the substrate. This adsorption happens fundamentally because of the attractive force between ions in the solution and surface of the substrate. **Fig 1.18** shows the schematic representation of the SILAR method. This forces maybe Van-der Wall's force or just the chemical attractive force. Therefore, they possess residual or unbalanced force and hold the substrate particles, and thus, the adsorbed atoms can be held on the surface and followed by reacting with anion resulting in the formation of thin films.

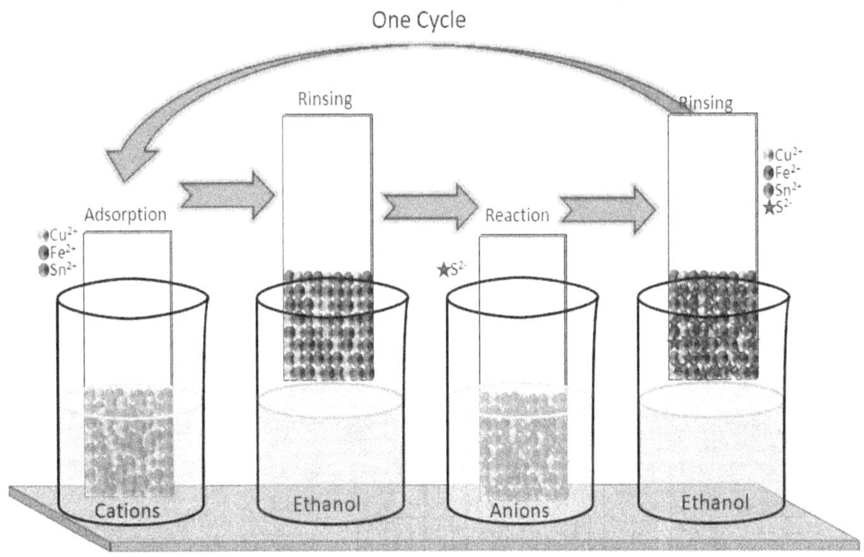

Figure 1.18 Schematic representation of SILAR method to deposit CFTS.

1.13.2 Hydrothermal synthesis

This method is used to synthesis various nanoparticles that depend upon the solubility of materials in hot water and under high pressure [74, 75]. The growth of these nanoparticles is performed in an apparatus consisting of steel which can resist high pressure vessel called an autoclave, in which the precursor is loaded along with the solvent, for example water. If the solvent is an aqueous solution, then this method is called hydrothermal, otherwise, it is known as solvothermal. Hydrothermal synthesis includes the various techniques of crystallizing substances from high-temperature aqueous solutions at high vapor pressure. The main advantage of this method is the growth of good quality crystalline nanoparticles with controlled size, morphology, and shape when compared to other methods. In other words, it is the ability to create a crystalline phase which is not stable at the melting point. There are three different types of hydrothermal methods such as temperature-difference method, temperature-reduction technique, and metastable-phase techniques. Among them, the temperature-difference method is the most extensively used one. However, the high cost and the inability to observe the growth of crystal limited the development of this method further extend.

Furthermore, this autoclave material must be inert to the solvents used. One of the most important elements of this autoclave is its closure, as it needs to withstand high pressure and temperature, the most famous Bridgman seal is used for its closure. This autoclave also must prevent the problem of corrosion of the internal cavities, hence protective inserts are used. This too will have the same shape as that of the autoclave and would fit in the internal cavity of the autoclave. These inserts can be of silver, gold, platinum, glass, or teflon depending on the temperature used.

1.13.3. Sol-gel

The sol-gel method is also one of the well-established synthetic approaches to prepare novel metal chalcogenides and metal oxide nanoparticles along with mixed oxide composites. The sol-gel process involves the conversion of monomers into a colloidal solution that acts as the precursor for an integrated network of either discrete particles or a network of polymers. The sol-gel method requires only a few steps to deliver the final output products with help of hydrolysis, condensation, and drying process. At first, the corresponding metal precursor undergoes rapid hydrolysis to form the metal hydroxide solution, followed by an immediate condensation which results in the formation of gels. Finally, the obtained gel is kept for the drying process and thus the resulting product is converted into Xerogel or Aerogel based on the mode of drying.

It is the chemical transformation of a system from a liquid "sol" into a gelatinous network "gel" phase with the subsequent transition into solid oxide material. Sol-gel methods can be classified into two types based on the nature of the solvents used such as aqueous sol-gel and non-aqueous sol-gel methods. In the former case, water is used as the reaction medium whereas, in the later one organic solvents are used. This process can provide better control of the whole reactions involved during the synthesis of solids. Furthermore, in the aqueous sol-gel method, oxygen is necessary for the formation of metal oxides, which is then supplied by the water solvents used. The schematic of the reaction pathway for the production of metal chalcogenides via the sol-gel method is shown in **Fig. 1.19**.

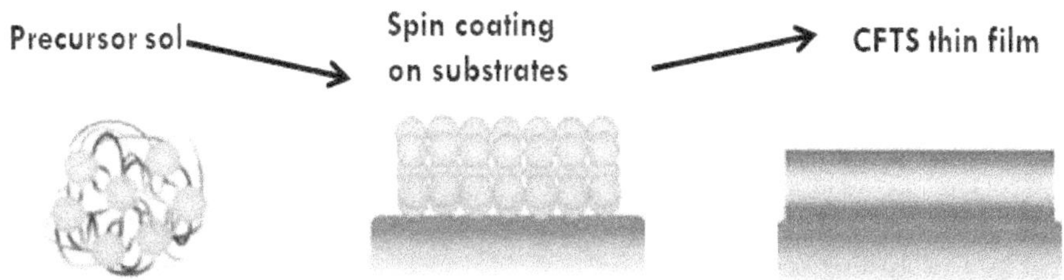

Figure 1.19 Synthesis of CFTS thin films via sol-gel method.

To control the size and shape of the sol-gel prepared materials, the addition of acid or bases catalyst is employed. This will be useful in the separation of hydrolysis and condensation stages in the process. It is also useful in the preparation of powder samples as the gels can break into granules during the drying procedure to produce granules of high surface area materials. Also, the removal of the solvents can be done by direct precipitation methods based on the addition of non-solvents. In this case, an aqueous solution is prepared and precipitated by dilution of this solution with a hydrophilic non-solvent. Furthermore, sol-gel methods are considered as easy procedures to modulate the final morphology of the materials. However, we know that the thermal treatment generally needs to be transforming the fried gel into the desired materials, but sometimes it leads to the uncontrolled growth of the nanocrystals. So, the tailoring of the exposed surface is not a simple issue that can achieve by this method.

1.14 Characterization analysis techniques

For structural analysis of these synthesized materials, there are large numbers of characterizations techniques used like X-ray diffraction (XRD), Scanning electron microscope (SEM), Field emission scanning electron microscope (FESEM), High-Resolution Transmission electron microscopy (HRTEM), Raman spectroscopy, energy dispersive X-ray analysis (EDX), Atomic force microscopy (AFM), etc. For understanding the properties of the deposited thin films, the characterization at both material and device stages are important. Since the arrangements of its composition and elemental distribution are not often homogeneously distributed, it may lead to distortions in the electrical properties and optical properties of the thin films. These are some of the main deposition techniques and characterization methods we are going to discuss briefly, since, in this thesis work, we have made use of these pioneering techniques in depth.

1.14.1 X-ray diffraction (XRD)

The X-ray diffraction (XRD) is a notable non-destructive technique for characterization that is widely used to analyze bulk, powder, and thin-film samples. It is a non-contact, non-destructive method and it provides the nature of crystallinity, phase, and orientation of the grains in the films. The structural identification and determination of lattice parameters are done based on the interpretation of the XRD patterns. The phenomenon of XRD consists of the reflection of X-rays from the different crystallographic planes. When an X-ray beam strikes the surface of the crystal at an angle θ, constructive interferences are observed. Diffraction occurs only when the wavelength of the wave motion is the same order of magnitude as the repeat distance between scattering centers. This condition of diffraction is known as Bragg's law and is given by the relation

$$2\ d\ \sin(\theta) = n\lambda \tag{1}$$

Where d is the inter-planar spacing (spacing between consecutive parallel planes), θ is the incident angle, n is the order of diffraction (n = 1), and λ is the wavelength of monochromatic X-ray.

The identification of phases is determined by comparing its diffraction pattern to know diffraction patterns using JCPD cards (Joint Committee on Powder Diffraction Standards card). The detailed information of XRD analysis has been mentioned in the literature [86, 87]. The absence of reflection peaks indicates that the samples are amorphous. Whereas, many reflection peaks indicates polycrystalline growth. To determine the average crystallite size in the case of polycrystalline samples is determined from Scherrer's formula [88]

$$D = \frac{K\,\lambda}{\beta\,\cos\theta} \tag{2}$$

Where D is the crystallite size, β is the Corrected FWHM of the most intense peak, θ is the Bragg's angle, K is the constant.

1.14.2 Raman spectroscopy

Raman spectroscopy is a spectroscopic technique used to study the scattering of light by the vibrating molecules and other low-frequency modes and to characterize the chemical composition and confirm the structure of the materials. Raman studies are done in this work to identify the secondary phases in the synthesized samples as a complementary technique for XRD. It is mainly based on the inelastic scattering of the monochromatic light, usually from a laser in the visible, near-infrared, or ultraviolet range. Typically, a sample is illuminated with a laser beam, this light is then sent through a monochromator. This light may be reflected, absorbed, or scattered. It is this scattering of the radiation that occurs which gives us the information regarding the molecular structure. There are two possibilities to observe vibrational mode peaks when the scattering of the incident light interact with the molecules of the materials:

- If the energy of the incident radiation is being elastically diffused without a change in the energy, then it is called Rayleigh scattering.
- If a small portion of light interacts with the material and the absorption of energy takes place, then it is Stokes radiation and if it gives energy to the photons incident on it, then it produces Anti-stokes radiation.

1.14.3 Field emission scanning electron microscopy (FE-SEM)

Field emission scanning electron microscopy (FESEM) provides topographical at a magnification as small as one nanometre (nm) to one billion of a millimeter (mm) on the surface of the materials and FESEM with EDX is necessary for elemental analysis. FESEM is a microscope that works with electrons instead of light. These electrons are liberated by the field emission source and the object is scanned by the electrons. Compared with the conventional scanning electron microscopy (SEM), field emission SEM produces clearer and less electrostatically distorted images.

Advantages of FE-SEM include:

- High quality and low voltage images with very less electrical charging of the samples.
- Reduced penetration of low kinetic energy electrons probes closer to the material surface.
- No need for any conducting coating required on the insulating materials.

- The ability to examine smaller-area contamination spots at electron accelerating voltages.

The image formation happens when the primary probe bombards the object secondary electrons are emitted from the object surface with a certain velocity. The secondary electrons which are attracted by the corona strike the scintillator that produces photons. The signal produced by the scintillator is amplified and transduced to a video signal that is fed to a cathode ray tube. Then this contrast in the real-time image that appears on the screen reflects the structure on the surface of the object. Parallel to these analog images, a digital image can be generated which can be further magnified or processed to form the required images.

1.14.4 High-resolution transmission electron microscopy (HR-TEM)

High-resolution transmission electron microscopy (HR-TEM) provides direct images of the atomic structure of the samples. Hence, it is an important tool to obtain direct information about the crystallographic structure of materials from images. It is also an ultimate tool in imaging defects as our thesis mainly deals with defects and grain boundary studies. It also shows directly a two-dimensional projection of the crystal with defects and all. High phase contrast images can be acquired as small as crystals. HRTEM images differ from the conventional electron microscopy because here the images are formed from the interference in the image plane. The electrons get interacted with the sample independently and the electron wave goes through the imaging system where a phase change occurs. The image recorded is not a direct representation of the sample structure. And also it is an invaluable tool for analysing the nanoscale properties of the materials due to the high resolution it offers.

1.14.5 X-ray photoelectron spectroscopy (XPS)

X-ray photoelectron spectroscopy (XPS) is a technique that is used for analyzing the surface chemistry of a material. XPS can be used to measure the elemental composition, empirical formula, chemical state, and electronic state of the elements within a material. It is the most widely used surface analysis technique as it can be applied to a wide range of materials. The average depth of analysis for an XPS measurement is approximately 5 nm. XPS spectra are obtained by irradiating a solid surface with a beam of X-rays while simultaneously measuring the kinetic energy of electrons that are emitted from the top of the material being analyzed. However, because XPS is an ultra-high vacuum technique, the sample to be analyzed has first to be evacuated. Physical Electronics XPS instruments function in a manner analogous to SEM/EDS instruments that use a finely focused electron beam to create SEM images for sample viewing and point spectra or images for compositional analysis.

XPS is typically accomplished by exciting a sample's surface with mono-energetic Al kα x-rays causing photoelectrons to be emitted from the sample surface. An electron energy analyzer is used to measure the energy of the emitted photoelectrons. From the binding energy and intensity of a photoelectron peak, the elemental identity, chemical state, and quantity of a detected element can be determined. By measuring the kinetic energy of the emitted electrons, it is possible to determine which elements are near a material's surface, their chemical states, and the binding energy of the electron. The binding energy depends upon several factors, including the following:

- The element from which the electron is emitted.
- The orbital from which the electron is ejected.
- The chemical environment of the atom from which the electron was emitted.

1.14.6 UV-Vis spectroscopy

The optical properties of the thin films and the powder samples were analyzed by UV-visible spectrophotometer in the range of 200-1000 nm wavelength range. Optical photon incident on any material may be reflected, transmitted, or absorbed. The phenomenon of radiation absorption in a material is due to the following activities:

- Inner shell electrons
- Valence band electrons
- Free carriers including electrons
- Electrons bound to localized impurity centers or defects

In semiconductors, the optical absorption spectra generally exhibit a sharp rise at certain values of the incident photon energy. This can be further attributed to the excitation of electrons from the valence band to the conduction band.

Direct and indirect are the two kinds of optical transitions that can occur in the crystalline semiconductor. Direct transition involves the interaction of an electromagnetic wave with an electron in the valence band which is excited across the fundamental gap to the conduction band. A vertical transition of electrons from the valence band to the conduction band occurs, such that there will be no change in the momentum of the electrons. In the indirect transition, it involves interaction with lattice vibration. The momentum change can be taken up by the phonon.

1.14.7 Atomic force microscopy (AFM)

Atomic force microscopy is arguably the most versatile and powerful microscopy technology for studying samples at the nanoscale. It is versatile because an atomic force microscope can not only image in three-dimensional topography, but it also provides various types of surface measurements to the needs of scientists and engineers. It is powerful because an AFM can generate images at atomic resolution with angstrom scale resolution height information, with minimum sample preparation.

AFM principle can be explained in 3 sections,

- **Surface sensing:** An AFM utilizes a cantilever with a sharp tip to look over an example surface. As the tip moves toward the surface, the short proximity, the attractive force between the surface and the tip makes the cantilever deflect towards the surface. Nonetheless, as the cantilever is brought considerably nearer to the surface, with the end goal that the tip makes contact with it, increasingly repulsive force dominates and makes the cantilever redirect away from the surface.

- **Detection method:** A laser beam is utilized to detect cantilever deflections towards or away from the surface. By reflecting an incident beam off the level top of the cantilever, any cantilever avoidance will cause slight alters in the course of the reflected shaft. A position-delicate photodiode (PSPD) can be utilized to follow these changes. The laser beam is used to detect cantilever deflections towards or away from the surface.

- **Imaging:** An AFM pictures the topography of a sample surface by scanning the cantilever over a region of interest. The hills and valley like highlights on the sample influence the deflection of the cantilever, which is observed by the PSPD. By utilizing a feedback loop to control the height of the tip over the surface consequently keeping up a steady laser position.

1.14.8 Solar simulator

A wide assortment of measurements is being utilized to decide the electrical characteristics of photovoltaic cells. We as a whole realize that when the cell is illuminated, optically created charge carriers produce an electric flow when the cell is associated with an external load. The measurements are normally done at various intensities of light and variable temperature conditions The current density-voltage (J-V) measurements are investigated to ascertain the increase in the current and voltage to calculate or determine the power conversion efficiency. The different losses are likewise decided from the electrical characterizations. Furthermore, it helps in manufacturing the gadget as productively as conceivable with insignificant losses.

To make these electrical measurements, the solar simulator Model Keithley 5460 source meter is used for calculating the J-V characterizations. The solar simulator also helps in the analysis of mathematical data with the help of software and graphics. This software helps in conducting the tests specifically on solar cells and derives the common photovoltaic cell parameters from the test data.

1.14.9 Photoelectrochemical (PEC) characterization

The above-mentioned characterization techniques are reliable but rather expensive and time-consuming. The preparative parameters of the quaternary chalcogenide semiconducting electrode by photoelectrochemical (PEC) method are one of the new and improved reliable and unique techniques in the field of thin films [62, 63]. PEC cell is a photocurrent generated device composed of a photoactive semiconductor electrode, electrolyte. When the light is irradiated at the interface of electrolyte and semiconductor with an energy level greater than the bandgap of the semiconductor, then the electron-hole pairs are generated.

The PEC device fabricated using this method can act as a photoelectrode, which can be completed by using suitable electrolyte and counter electrodes. These are then illuminated with the help of appropriate light energy. The short circuit current (I_{sc}) and open-circuit voltage (V_{oc}) are the two initial parameters that are used for optimization and calculation of the power conversion efficiency. Finally, it can be observed that the values of optimized parameters obtained from the PEC measurements are well-matched with the optimized values generated by employing other characterization techniques.

1.15 Scope and objectives of the thesis

The literature review introduced above shows that the importance of partial substitution of Fe and Ba to replace Zn cations to handle antisite defects and to use in different multifunctional applications. Among these substitutions, Zn with Ba demonstrates the highest photocurrent for the photoelectrochemical cells. Additionally, the cation trade with different transition elements (Mn, Co, Fe, Ni) may present deep trap states or recombination centers as a result of their multiple charge states. Hence, replacing Zn with the element of least proximity in the periodic table isn't adequate to decrease the cationic disorder in the structure. All things considered, if the Zn is replaced with the far-off element in the periodic table like Ba whose electronic environment is different from Zn, maybe the cation-cation disorder could be avoided to a larger extent.

The earth-abundant quaternary chalcogenides exhibit attractive optical and electrical properties. These semiconducting nanoparticles have superior charge transport capability and optical properties making it suitable for various electronic and optoelectronic applications. If the synthesized particles are of nanoscale range, then these absorber materials can be used for various novel applications. These applications include energy-related and memory-related applications. The energy-related applications consist of photodetectors, solar cells, photoelectrochemical (PEC) cells for hydrogen evolution i.e. water splitting. Whereas, memory-related applications include resistive random access memory (RRAM), negative differential resistance (NDR) for bistable memory switching applications, etc.

Keeping these things in mind, the importance of the above-mentioned light absorber materials, the studies reported in this thesis are categorized as follows:

So, the main objectives of the thesis are:

- Development of cost-effective solution based p-type CFTS absorber layer for various multifunctional properties.
- Introducing various stabilizers and additives for eliminating defects and secondary phases.
- Structural, morphological, photo-physical, optical, and electrical investigation of the device for understanding the defect physics.

1.16 Organization of thesis

The present thesis is summarized into 5 chapters in the following manner: **Chapter 1** gives a brief introduction to the advantages of absorber layers like CFTS and CBTS materials over CZTS, CIGS, and CdTe absorber materials. Also, this chapter describes various characterization techniques. Finally, this thesis discusses about various multifunctional applications like memory devices, super-capacitors, photodetectors, and photoelectrochemical cells, and the effect of sulfurization are also described. This chapter concludes with the scope and objectives of this thesis. **Chapter 2** describes the synthesis of CFTS nanoparticles with the help of a solvothermal route using PEG with different chain lengths as solvent followed by sulfurization. Also, this chapter focuses on the photo-enhanced supercapacitive behavior of photoactive (CFTS) nanoparticles. **Chapter 3** deals with the fabrication of different types of CFTS thin films with the help of the sol-gel method and the spin coating process. Then these CFTS thin films were administered with different types of stabilizers to check whether this helps in reducing the secondary phase followed by the sulfurization process as the formation of the required phase wasn't completed. In **Chapter 4,** CFTS thin film was fabricated by a solution-based SILAR method which resulted in a pure stannite phase. These CFTS thin film was used for the memory applications. Finally, an n-type Bi_2S_3 thin film was fabricated beside the p-type CFTS thin film which then completes a p-n junction. **Chapter 5** describes the partial substitution of Ba in the CFTS compound semiconductor and studied the physical, chemical, and electrical properties of polycrystalline $Cu_2Fe_{1-x}Ba_xSnS_4$ (CFBTS) alloys with the help of the SILAR method without any post-annealing process. High performance photoelectrochemical (PEC) cell with Glass/FTO/CFBTS structure with improved stability was fabricated without the addition of any bilayers on top of CFBTS thin films. Finally, we have included the concluding remarks or inferences drawn from the entire research work presented in this thesis are summarized with future directions in **Chapter 6.**

Chapter 2

Photo Enhanced Supercapacitive Behaviour of Photoactive Cu_2FeSnS_4 (CFTS) Nanoparticles

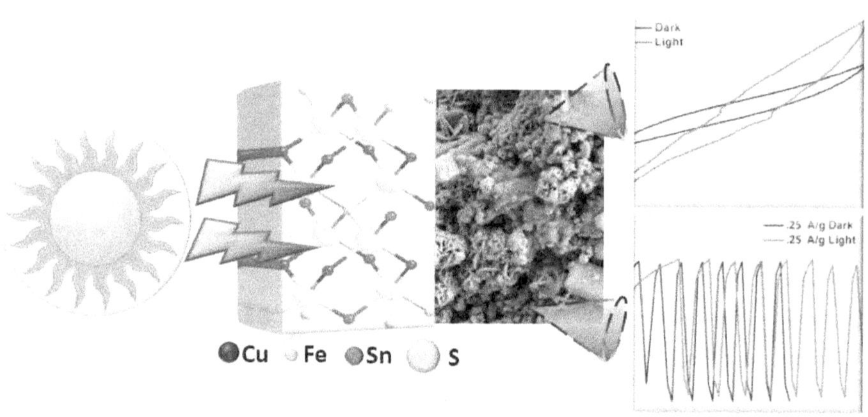

CHAPTER 2

Photo Enhanced Supercapacitive Behaviour of Photoactive Cu2FeSnS4 (CFTS) Nanoparticles

2.1 Introduction

CFTS can be utilized in various multifunctional applications as mentioned in the previous chapter. There are various applications related to energy storage and conversion and in this chapter, we are focusing on the storage of energy in the form of supercapacitors [89-91]. Also, would like to study the effect of photons with the help of photo-enhanced photo chargeable supercapacitors. The photo supercapacitors are new energy devices that combine the photoelectric conversion function of a solar cell with the energy storage capability [92-95]. Energy storage devices like supercapacitors are an essential element in modern technology. Among different types of energy storage devices available in the market, supercapacitors have attracted research and industries due to their advantages such as high power density, low cost, cycle-life, stability, etc [96].

Metal sulfides such as ternary/quaternary are considered an interesting class of materials as it can be used for multifunctional applications [97, 98]. Tang et al. reported the hydrothermal synthesis of pure CZTS delivering a specific capacitance of 138 F/g and was then enhanced to 591 F/g by the addition of reduced graphene oxide (rGO) [99]. Cations and anions are replaced with extrinsic impurities such as Mn, Fe, Co, Cd, Ni, and Se in CZTS to control the formation of the secondary phase and to study the influence on optical, electrical transport along with photovoltaic device performance characteristics [61,100-104]. There are several techniques involved in the synthesis of CFTS, which include solution-based depositions like chemical methods, successive ionic layer absorption and reaction (SILAR), solvothermal, etc [48, 105-107]. However, the majority of the fabrication process involves toxic elements and expensive steps and thus, it remains a huge challenge for the researchers to deploy it in practical applications on a large scale. Also, the solvents used for these reactions contain flammable, volatile, toxic elements that bring environmental pollution factors and will affect the sustainable development of the scientific society. There are many potential advantages of replacing organic compounds with aqueous or non-aqueous based solutions. of the main aqueous solvent used is polyethylene glycol (PEG) solutions.

Polyethylene glycol (HO–(CH$_2$CH$_2$O)$_n$–H) is available in a variety of molecular weights starting from 200 to tens of thousands. PEGs are colorless, odorless, and non-toxic stable at room temperature. These are also considered inert and do not react readily with other materials. Unlike several of the other solvents such as ionic liquids, where the level of toxicity and environmental problems are unknown, whereas, for PEG a complete list of toxicity profiles is available for a wide range of molecular weights of PEG [108, 109]. One report on supercapacitors suggests that electrodes are fabricated with conducting polymer, carbon-based materials [110, 111], and metal chalcogenide and oxides like nickel cobaltite (NiCo$_2$O$_4$) [112], MnO$_2$ [113], Al$_2$O$_3$ [111]. On the other hand, metal chalcogenides are also used as photo absorbing materials that have broad scope in studying the photoinduced capacitance enhanced behavior of the materials. A couple of research attempt was done in the solar cells with energy storage devices to develop photo-supercapacitors [114-116]. Many of these studies were based on the separate integration of the energy sources with storage devices, but there are some efforts on combining both energy conversion and storage into a single and compact electrochemical cell [117, 118]. In addition, the effects of solar light in supercapacitor application may provide a novel approach for the multifunctional energy storage supercapacitor devices [119].

In this chapter, we state a simple and mild solvothermal reaction to synthesize CFTS nanoparticles and sulfurized CFTS nanoparticles at 500 $^{\circ}$C. Different polyethylene glycol (PEG) of varying chain lengths was used as a solvent to study the effect on structural, morphological, and photoconductive properties of the as-prepared CFTS nanoparticles. The present study also exploits the electrochemical performances of the as-prepared CFTS nanoparticles and sulfurized CFTS nanoparticles. The synthesized CFTS demonstrated a significant increase in the specific capacitance under light illumination as compared to the capacitance in the dark condition. And the sulfurized samples showed less performance compared with as-prepared CFTS nanoparticles due to the presence of secondary phases such as Cu$_2$S and Sn$_2$S$_3$. Moreover, in presence of light, it enables the conversion of the photochemical energy into electrochemical energy.

2.2. Experiment

2.2.1. Chemicals and materials required

Copper nitrate (99% Thomas Baker), Iron sulfate (99% Thomas Baker), Tin chloride (99% Thomas Baker), and Thiourea (99% Thomas Baker) were the precursors used for the synthesis of the CFTS nanoparticles and polyethylene glycol (PEG) (Loba Chemie Pvt. Ltd.) with three different molecular weights (PEG-200, PEG-400, and PEG-600) were used as the solvents. Ethanol (Analytical reagent) and Deionised water (DI water) were used for washing CFTS nanoparticles. All these chemicals are commercially available and used without further purification.

2.2.2 Synthesis of CFTS nanoparticles by solvothermal method

Here CFTS nanoparticles were synthesized by solvothermal route using PEG as the main solvent. Copper nitrate (0.02 M), Iron sulfate (0.01 M), Tin chloride (0.01 M), and Thiourea (0.1 M) were dissolved in 20 ml of PEG and stirred for 5 hrs in a glass beaker. the entire reaction mixture was poured into a Teflon beaker to undergo the solvothermal reaction conducted at 200 °C for 24 hrs. After the reaction, it was allowed to cool down to room temperature and CFTS nanoparticles were obtained by washing it with DI water and ethanol followed by centrifugation. Finally, the sediments undergo further drying to form powder as shown in **Fig. 2.1**. To study the effect of sulfurization all three samples were annealed in presence of sulfur powder in the flow of inert gas (Argon) atmosphere. The as-prepared CFTS nanoparticles using PEG-200, PEG-400, and PEG-600 will be labelled as CFTS1, CFTS2, and CFTS3 and the sulfurized CFTS nanoparticles will be labelled as CFTS S1, CFTS S2, and CFTS S3 respectively.

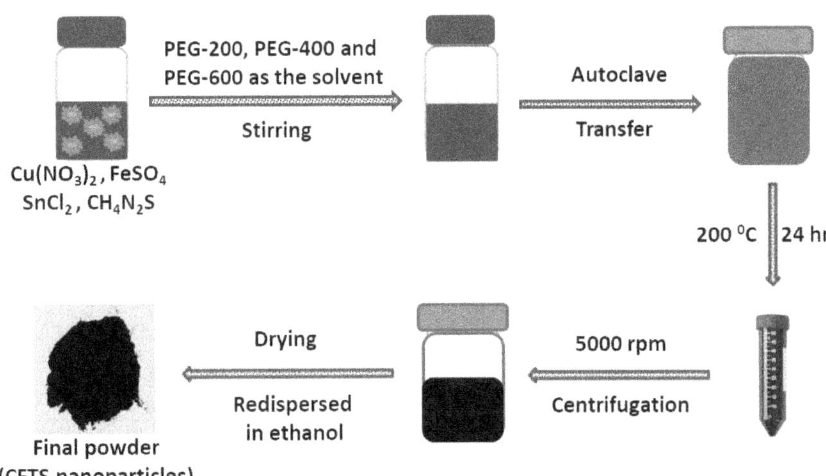

Figure 2.1 Schematic illustration of the procedure for the synthesis of Cu$_2$FeSnS$_4$ nanoparticles.

2.2.3 Characterization of materials

X-ray diffraction (XRD) was used to investigate the crystal structure and phase purity of all the CFTS nanoparticles. Raman spectroscopy was done using an excitation laser of 633 nm wavelengths and power of 0.6 mW for fluorescence suppression and also high energy of UV photons can damage the sample. All the CFTS nanoparticles were characterized using the high-resolution transmission electron microscope (HRTEM) images by drop-casting the nanoparticles on a carbon-coated Cu grid. The morphological analysis of the CFTS nanoparticles was studied using a field emission scanning electron microscope (FESEM). UV-Vis spectrophotometry was done for obtaining the absorption spectrum. The $I-V$ characteristics of these samples were measured using Keithley digital source meter under an applied potential scan between two lateral Al (thermally deposited on a glass slide) electrode with active materials composed of carbon and CFTS nanocrystals. The electrochemical analysis using cyclic voltammetry (CV), galvanostatic charge/discharge (GCD) measurements, and electrochemical impedance spectroscopy (EIS) were investigated by using the Potentiostat/Galvanostat/EIS instrument. Room temperature electrochemical measurement of CFTS2 and CFTS S2 was carried out in a three-electrode system with 1 M KCl aqueous electrolyte, Ag/AgCl as a reference electrode, and a platinum wire as the counter electrode. The schematic diagram of the three-electrode system for the measurement of photoenehanced supercapacitor is shown in the Fig. 2.2.

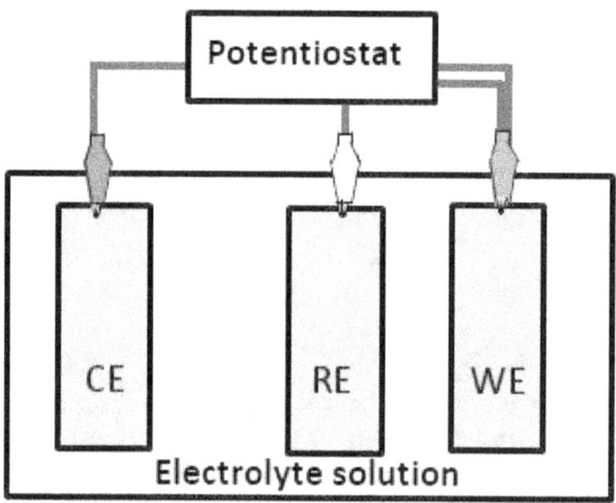

Figure 2.2 Schematic diagram of three-electrode system.

2.3 Results and discussion

2.3.1 As-prepared CFTS nanoparticles

(a) X-ray diffraction

The XRD patterns of as-prepared CFTS1, CFTS2, and CFTS3 are shown in **Fig. 2.3**. The as-prepared CFTS1, CFTS2, and CFTS3 can be indexed to the stannite phase of CFTS with no other extra impurity peaks. The major peaks observed at 29.18°, 48.06°, and 56.82° correspond to (112), (204), and (312) planes respectively (JCPDS card no. 74-1025). The XRD alone is not possible to confirm the phase purity of CFTS, so Raman spectroscopy was done to study the phase purity of the samples.

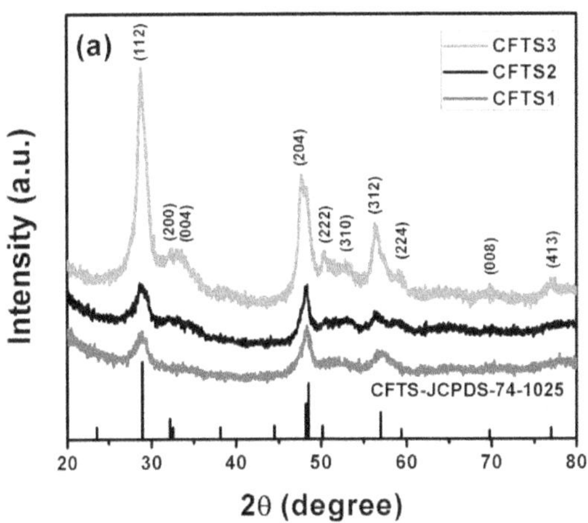

Figure 2.3 XRD patterns of as-synthesized CFTS1, CFT2 and CFTS3 nanoparticles.

(b) Raman spectroscopy

As the XRD is not enough to confirm the phase purity and thus we studied the Raman spectroscopy to identify whether any secondary peaks present. A peak at 320 cm^{-1} confirmed the presence of the stannite phase of CFTS1, CFTS2, and CFTS3 as shown in **Fig. 2.4**. The identified peak is of small intensity and wide, probably because of the poor crystallinity of the as-prepared CFTS nanoparticles. Also, the formation of CFTS was confirmed by the presence of the peak (320 cm^{-1}), suggesting strong asymmetric vibration of a sulfur atom around tin [106]. However, no other impurity phases (e.g., Cu$_x$S and FeS) were present in the

CFTS nanoparticles which further confirmed the absence of other secondary phases for the as-prepared CFTS samples.

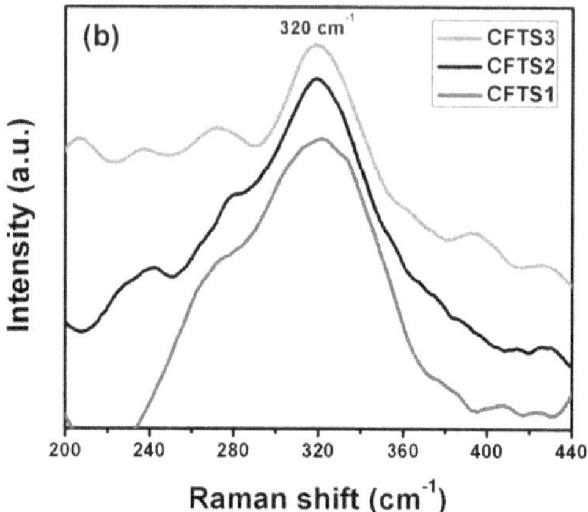

Figure 2.4 Raman spectra of as-synthesized CFTS1, CFT2 and CFTS3 nanoparticles.

(c) UV-Vis spectroscopy

The optical properties of all the as-prepared CFTS nanoparticles were analyzed by the UV-Vis absorption spectrophotometer and is shown in **Fig. 2.5**. The liquid phase suspensions were prepared by dissolving 1 mg of the CFTS nanoparticles in 3 ml of ethanol. The bandgap of the nanoparticles was estimated by extrapolating the linear region of the plot (αhν)2 vs. (hν) to intercept on the horizontal photon energy axis as illustrated in **Fig. 2.5(b)**. The obtained band gap values for CFTS nanoparticles synthesized from PEG-200, PEG-400 and PEG-600 were 1.39 eV (CFTS1), 1.41 eV (CFTS2) and 1.42 eV (CFTS3) respectively. During synthesis, it was observed that the particle size increases along with using long-chain polyethylene glycol. Interestingly the shift in the bandgap is related to the chain length of the PEG used. As the chain length increases, the band gap also increases for the CFTS nanoparticles, the probably smaller chain length of PEG developed fewer defects on the structures. The changes in the band gaps were due to the presence of surface defects on the CFTS nanoparticles. Owing to these defects, the absorption edge in the conduction band has shifted to the higher energy states. As a result of this, the energy states near the conduction band will be crowded and thus the value of the optical band gap has increased. Also, the obtained band gap values were found to be in good agreement with the reported results of CFTS [116].

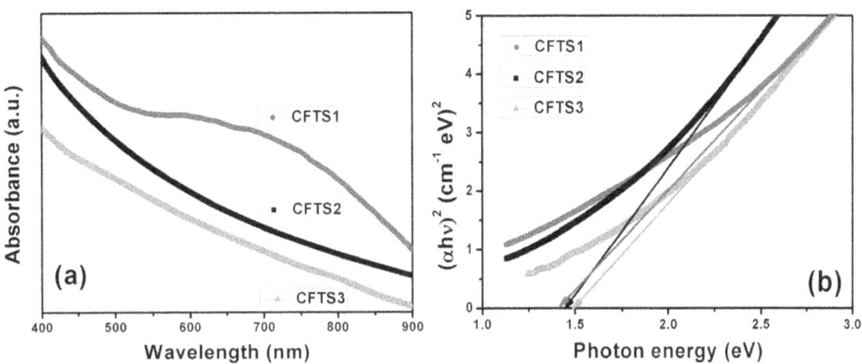

Figure 2.5 (a) Optical absorptionspectra, and **(b)** corresponding plots of hv vs (αhv)^2 to determine the optical band gap of CFTS1, CFTS2 and CFTS3 nanoparticles.

(d) High-resolution transmission electron microscopy (HRTEM)

HRTEM was employed to identify the shape, size, and particle distribution of the as-synthesized CFTS nanoparticles. **Fig. 2.6(a-c)** shows the HRTEM images and **Fig. 2.6(d-f)** shows the selected area electron diffraction (SAED) of the as-synthesized CFTS nanoparticles.

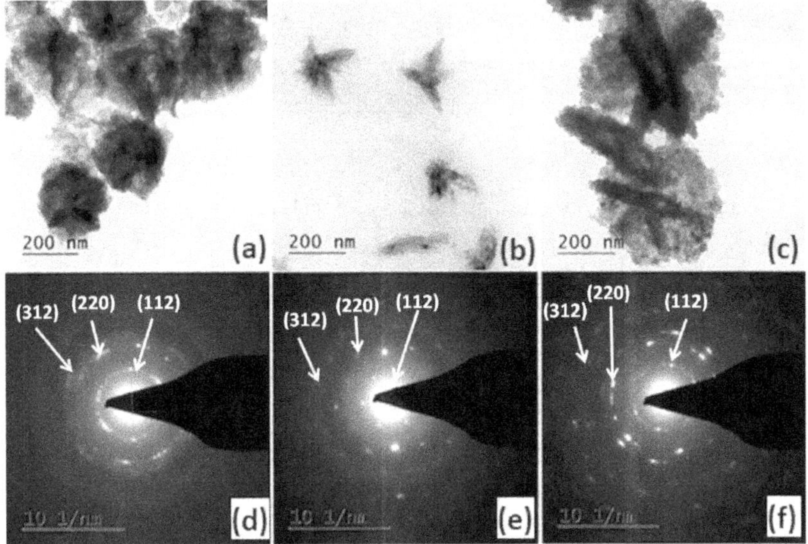

Figure 2.6 (a, b, c) The HR-TEM images and **(d, e, f)** SAED patterns of CFTS1, CFTS2 and CFTS3 nanoparticles.

It shows that the CFTS1 and CFTS3 nanoparticles are spherical in shape and have a diameter of 200 nm and 400 nm respectively, whereas CFTS2 has thorn-like projections in five directions which is evident from TEM images. The SAED pattern shows the polycrystalline nature of the samples and concentric circles for both CFTS1 and CFTS3 can be linked to (111), (220), and (311) planes and corresponds to the stannite crystal structure of CFTS. Also, this observation and plane indexing agree well with XRD. While the SAED pattern of CFTS2 reveals its monocrystalline nature.

(e) X-ray photoelectron spectroscopy (XPS)

X-ray photoelectron spectroscopy (XPS) was used to study the surface composition and the valence states of elements in the as-synthesized CFTS nanoparticles. **Fig. 2.7** shows the Cu $2p$, Fe $2p$, Sn $3d$, and S $2p$ high-resolution XPS peaks of the CFTS nanoparticles. The binding energies of Cu, Fe, Sn, and S were consistent with the reported literature [78, 106, 120, 121]. Figure 2.6(a) shows the most intense binding peaks for Cu $2p_{3/2}$ and Cu $2p_{1/2}$ were located at 932.3 and 952.1 eV respectively with a peak splitting of 19.8 eV and it indicates the presence of Cu in +1 oxidation state [78,121]. The Fe 2p peaks had very low intensity with 2p3/2 peak at 710.2 eV as a shoulder closer to the Sn 3p3/2 peak. The peak position indicates the oxidation state of Fe is +2 as shown in **Fig. 2.7(b)** [106, 122]. The Sn $3d_{5/2}$ and $3d_{3/2}$ peaks were present at 486.5 and 495.0 eV respectively with a peak splitting of 8.5 eV suggests Sn with +4 oxidation state (**Fig. 2.7(c)**) [119, 106]. Finally, **Fig. 2.7(d)** shows the high-resolution spectra of sulfur, in which it indicates peaks of $2p_{3/2}$ and $2p_{1/2}$ of S^{2-}, S_2^{2-} and $(SO_4)^{2-}$ are observed [123, 124].

Detailed fitting is presented in **Fig. 2.8** that indicates sulfur in -2 state and the concentration of sulfur is relatively higher on the surface of CFTS2. XPS curve fitting also known as peak fitting has been widely used for extracting chemical information from the overlapping features in the high resolution XPS spectra. Non-linear least squares (NLLS) is the method used and it is considered to be a superior method of curve fitting in XPS spectra, particularly when the fitted lineshape is complex. And the purpose of this curve fitting with a set of component peaks is to separate the photoemission signal originating from the distinct elemental or chemical states. Peaks observed at 168.3 and 169.2 eV indicate that the CFTS nanoparticles are partially oxidized [106, 121, 122]. The oxidation states of these CFTS nanoparticles matches with the chemical formula $(Cu_2^{+1}Fe^{+2}Sn^{+4}S_4^{-2})$.

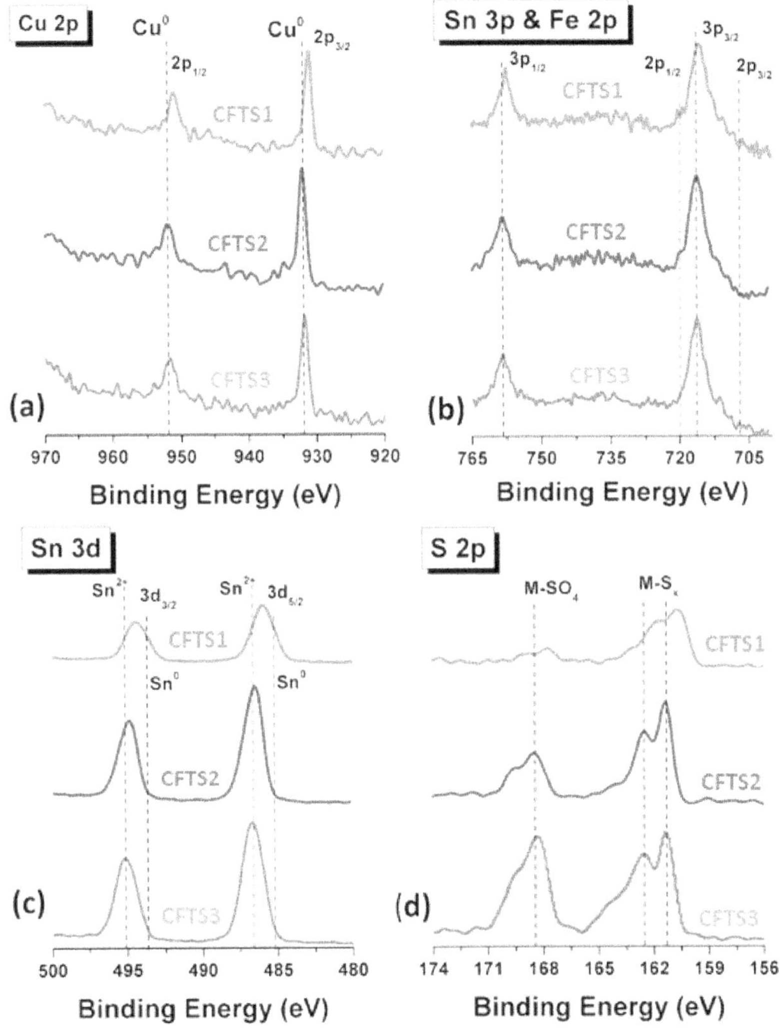

Figure 2.7 High resolution XPS spectra of **(a)** Cu 2*p*, **(b)** Fe 2*p*, **(c)** Sn 3*d* and **(d)** S 2*p* of CFTS1, CFTS2 and CFTS3 nanoparticles.

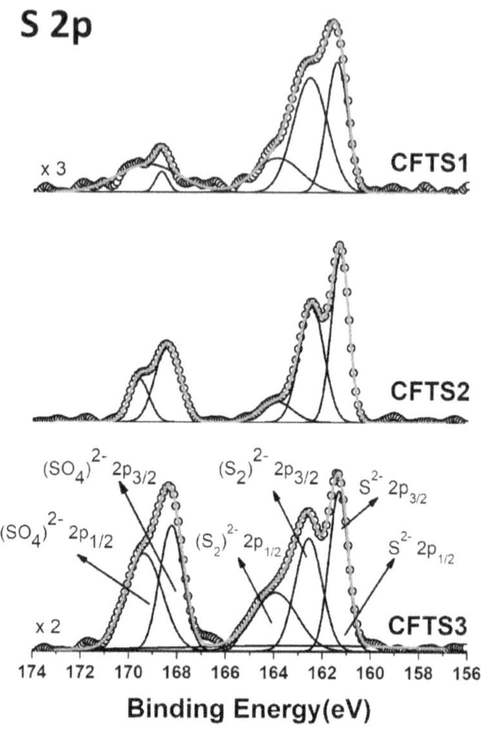

Figure 2.8 Fitting for high resolution XPS spectra of S 2p for CFTS1, CFTS2 and CFTS3 nanoparticles.

(f) Field emission scanning electron microscopy (FE-SEM)

The surface morphology of the CFTS nanoparticles is studied with the help of low and high magnification SEM images as shown in **Fig. 2.9**. It was observed that CFTS1 exhibited regular plate-like structures combining to form a spherical shape with an average size of 200 nm which is evident from the high magnification image (**Fig. 2.9(d)**), which is consistent with the TEM results. Whereas, CFTS2 shows irregular shapes, consisting of spherical and aggregated particles which is clear from the high magnification images (**Fig. 2.9 (b)** and **Fig. 2.9 (e)**) along with few projections. Finally, CFTS3 nanoparticles show the presence of a majority of plate-like and little spherical morphological appearance. This type of morphology evolution is due to the Ostwald ripening mechanism. The crystal nucleus formation takes place in the reaction solution which further grows into large nanoparticles through the Ostwald ripening mechanism. As and when the chain length of PEG decreases the crystal nucleus grows into spherical sheet like structure along a two-dimensional axis through the Ostwald ripening process. Thus large sphere-like particles are formed with the self-assembly of many sheet-like particles. Additional FE-SEM and HR-TEM images of CFTS1, CFTS2, and CFTS3 nanoparticles are also shown in **Fig. 2.10**.

Figure 2.9 Low and High magnification of the FE-SEM images of **(a, d)** CFTS1, **(b, e)** CFTS2 and **(c, f)** CFTS3 nanoparticles.

Figure 2.10 Additional HR-TEM and FE-SEM images of **(a,b)** CFTS1, **(c,d)** CFTS2 and **(e,f)** CFTS3 nanoparticles.

(g) *I-V* **characteristics of CFTS nanoparticles**

To investigate the potential application of the CFTS nanoparticles in photovoltaics, we have studied the photo-conductivity of these quaternary chalcogenide nanoparticles from *I-V* characteristics. **Fig. 2.11 (a-c)** shows the dark and under 100 mW/ cm^{-2} light illumination (xenon lamp) for CFTS1, CFTS2 and CFTS3 thin films. The current value of these films increases when irradiated with light. CFTS1 film delivers the maximum current under this illumination followed by CFTS2 and CFTS3. As can be seen that CFTS1 delivers 1.2 mA followed by 1.1 mA and 0.6 mA for CFTS2 and CFTS3 respectively under light illumination. This increase in the conductivity can be attributed to the decrease in the chain length of polyethylene glycol induced regular arrangement of the particles with good inter-particle electrical contact as seen from the SEM images. Also, as the chain length of PEG increases, the current density decreases for these films.

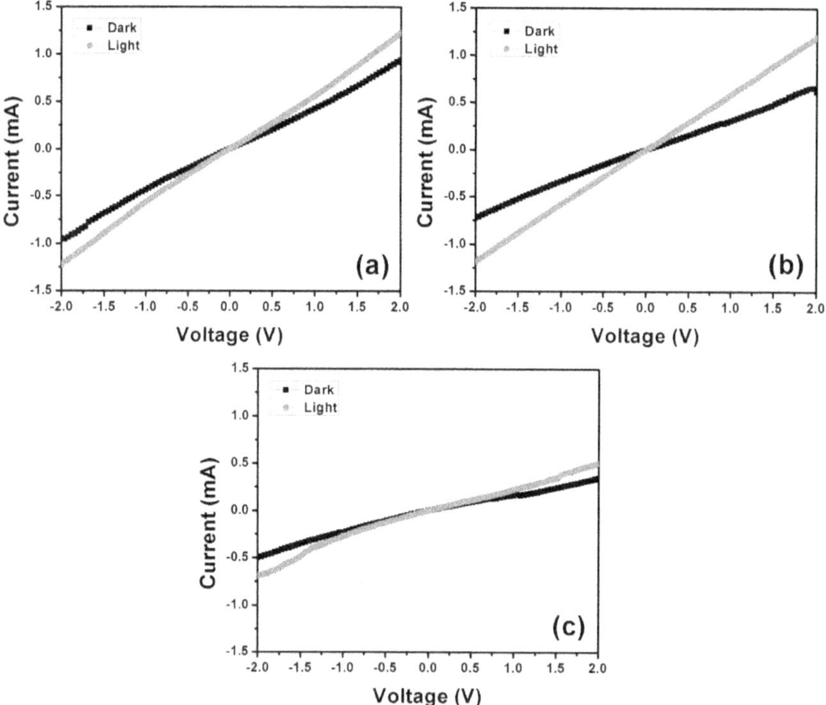

Figure 2.11 *I-V* characteristics of **(a)** CFTS1, **(b)** CFTS2 and **(c)** CFTS3 nanoparticles in dark and under light illumination.

However, the higher current is exhibited by PEG with less chain length, whereas higher light to dark ratio is shown by CFTS2 nanoparticles. As CFTS2 nanoparticles contain comparatively larger grains along with large spherical and aggregated particles without any noticeable voids. The reduction in the recombination rate of the photo-generated charge carriers is due to the larger grains that have fewer grain boundaries. And thus CFTS2 nanoparticles are exhibiting the highest light to dark current. Hence, all other electrochemical performances are performed for CFTS2 nanoparticles.

(h) Cyclic voltammetry (CV)

Cyclic voltammetry (CV) is an electrochemical technique which measures the current response of a redox active solution to a potential sweep between a set of values. CV can be used to measure the specific capacitance (C) of an active material using the equation (1),

$$C = {}^{A}\!/_{2mVv} \tag{1}$$

Where, '**C**' is the specific capacitance, '**A**' is the absolute area, '**m**' is mass, '**V**' is voltage and 'v' is the scan rate in V/s.

Fig. 2.12 illustrates the CV curves recorded in both dark and under light conditions of the as-prepared CFTS2 nanoparticle at a scan rate of 50 mV/s and the remaining scan rates are shown below.

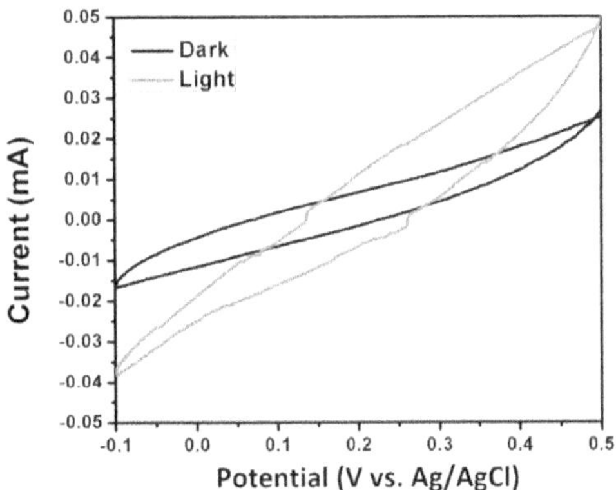

Figure 2.12 CV curves of as-prepared CFTS2 nanoparticles at 50 mV/s scan rate in dark and under light illumination.

The curves were measured under varying potential ranging from -0.1 to 0.5 V with a scanning speed of 5, 20, 50, 80, and 100 mV/s under dark and light illuminated conditions as shown in **Fig. 2.13 (a)** and **Fig. 2.13 (b)** for the as-prepared CFTS2 nanoparticles. At lower scan rates, all the samples showed a similar symmetry with good electrical double layer capacitor (EDLC) characteristics. From the comparison of both, the area under the curve for the dark and light conditions, the light illuminated sample showed larger capacitance when compared to the dark condition for CFTS nanoparticles. Among the three samples, light illuminated CFTS2 shows higher electrochemical specific capacitance when compared to the dark condition.

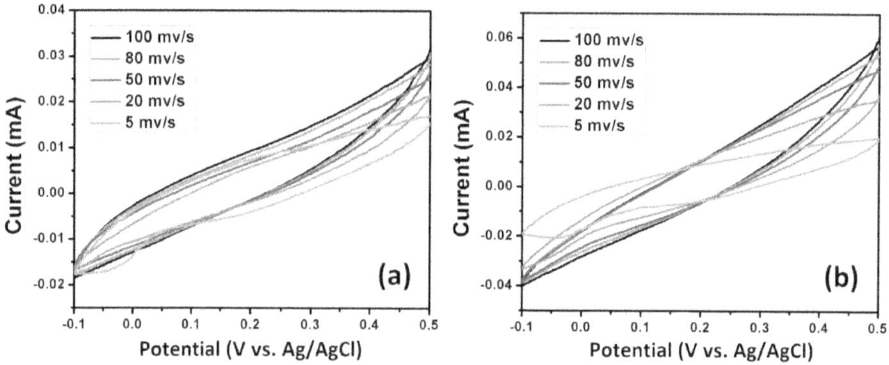

Figure 2.13 CV curves of CFTS2 nanoparticles at various scanning rates (5, 20, 50, 80, 100 mV/s) in **(a)** dark and **(b)** under light illumination.

Fig. 2.14 summarised the curves of specific capacitances with varying scan rates of as-prepared CFTS2 nanoparticles. In general, on increasing the scan rate the capacitance of the electrode is reduced. This is because, greater the scan rate faster will be the rate of charging and discharging, larger the resistance on the ion transport leading to the difficulty in penetration of ions deep into the pores and thereby reducing in capacitance of the electrode. These results indicate that, in the light illuminated conditions to provide greater advantages inthe electrode-electrolyte interface, there will be more interface in the generation of photon excited electrons and holes. The capacitance was enhanced after light illumination as in the case for as-prepared CFTS nanoparticles.

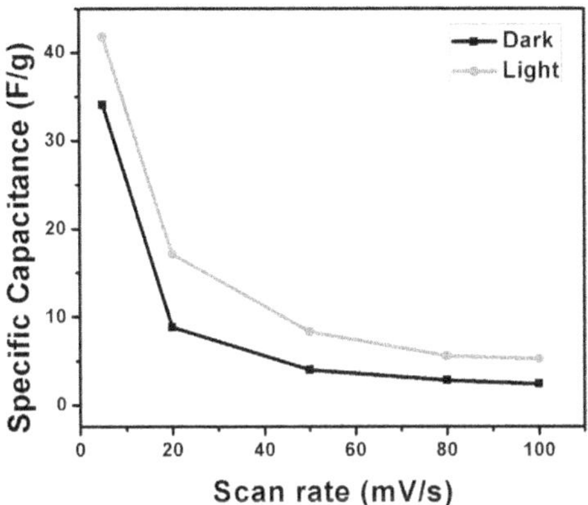

Figure 2.14 Comparative plots of specific capacitance as a function of scan rates (5, 20, 50, 80, 100 mV/s) in dark and under light illumination for as-prepared CFTS2 sample.

(i) Galvanostatic charge/discharge (GCD)

The comparison of galvanostatic charge/discharge (GCD) performances of as-prepared CFTS2 under light and dark conditions are shown in **Fig. 2.15**. The curves well indicated the EDLC behaviour of the material and the discharge curve of CFTS2 in dark condition for the first cycle exhibits a discharge time of 4 s, whereas when the light is applied for the same measurement, it exhibited an increase in the discharge time (15 s) which reveals the enhancement in the specific capacitance.

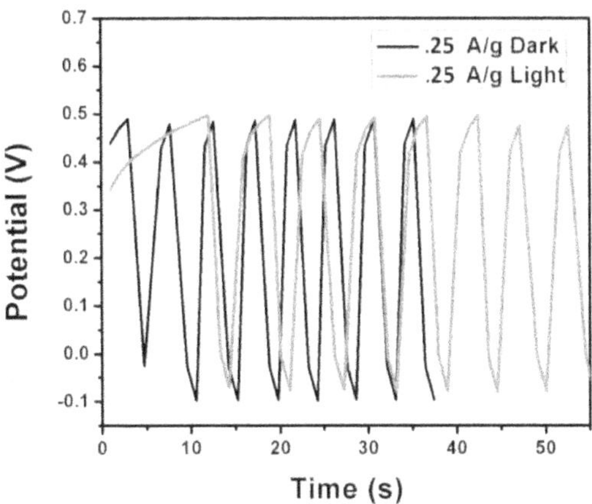

Figure 2.15 Galvanostatic charge/discharge profiles of as-prepared CFTS2 nanoparticles at a constant current density of 0.25 A/g in dark and under light illumination respectively.

The difference in the charge-discharge cycles of light to the dark condition was around 18 s. The increase in the charge-discharge time suggests the enhancement of capacitance in the light irradiation. GCD measurement also reveals that the CFTS2 sample is the best performing photo-enhanced super capacitive material.

(j) Electrochemical impedance spectroscopy (EIS)

Electrochemical impedance spectroscopy (EIS) was done to understand the electrochemical behaviour of these nanoparticles. ZSimpWin software was used to analyse the impedance parameters from a least-squares fit to an equivalent circuit. The EIS Nyquist plots (**Fig. 2.16**) exhibits two distinct parts, a semi-circle is observed in the high-frequency region indicates the effect of grain and followed by a straight line in the low-frequency region indicates the grain boundary effect. The intercept of these semicircles on the real axis gives the resistance of the grain (Rb) and grain boundary (Rgb) of the corresponding component which contributes to the impedance of the sample. And the capacitance values calculated from the grain and grain boundaries were 5.05 µF and 10.10 µF for dark condition and 17.31 µF and 25.37 µF under light illumination respectively. According to the EIS data, it can be seen that the resistance of the CFTS2 nanoparticles is decreasing under the light illumination condition. Lower interfacial charge transfer resistance is exhibited under light illumination due to which a superior capacitive behavior is observed than the dark condition.

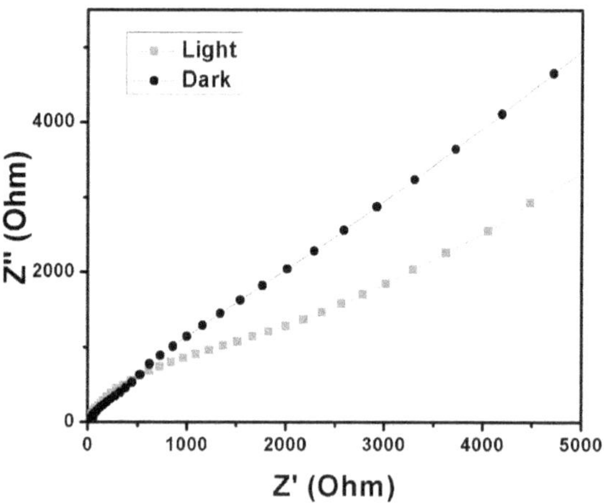

Figure 2.16 Nyquist plots of the CFTS2 nanoparticles in dark and under light illumination as obtained from the EIS study.

2.3.2. Sulfurized CFTS nanoparticles

(a) X-ray diffraction

The XRD patterns of as-prepared CFTS S1, CFTS S2, and CFTS S3 are shown in **Fig. 2.17**. Most of the peaks obtained from the sulfurized CFTS S1, CFTS S2, and CFTS S3 nanoparticles are matching well with the reported stannite phase according to the JCPDS card no. 74-1025 with few impurity peaks. These samples show a couple of secondary phases such as Sn_2S_3 (111) and Cu_2S (322) as confirmed from the JCPDS card nos. 14-0619 and 33-0490 respectively.

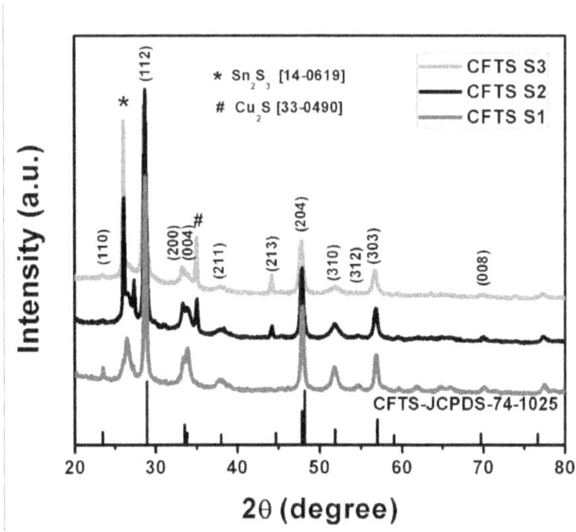

Figure 2.17 XRD patterns of sulfurized CFTS S1, CFTS S2 and CFTS S3 nanoparticles.

(b) Raman spectroscopy

The Raman spectroscopy for the sulfurized samples are shown in Figure 2.17 for CFTS S1, CFTS S2 and CFTS S3 exhibited the most prominent peak at 317 cm⁻¹ and other small peaks at 283 cm⁻¹ and 353 cm⁻¹ matched well with the stannite phase, also the prominent peak had a slight shift from the as-prepared samples (320 cm⁻¹) as shown in **Fig. 2.18**. This shift of Raman peak towards lower frequencies is either due to the existence of highly disorder distribution of cations. And this may be related to slight change in the length of the metal-S bond or due to the two kinds of anionic mode of S anions around Sn cations.

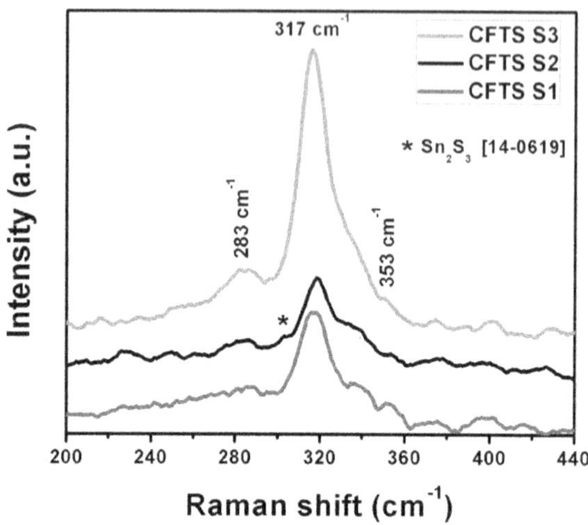

Figure 2.18 Raman spectra of sulfurized CFTS S1, CFTS S2 and CFTS S3 nanoparticles with a secondary phase "*" Sn_2S_3.

(c) UV-Vis spectroscopy

The optical properties for all the sulfurized CFTS nanoparticles were also analysed by the UV-Vis absorption spectrophotometer as shown in **Fig. 2.19**. The optical properties of sulfurized CFTS nanoparticles were also studied, CFTS S1, CFTS S2, and CFTS S3 samples also exhibit absorption in a wide spectrum range as seen in **Fig. 2.19(a)**.

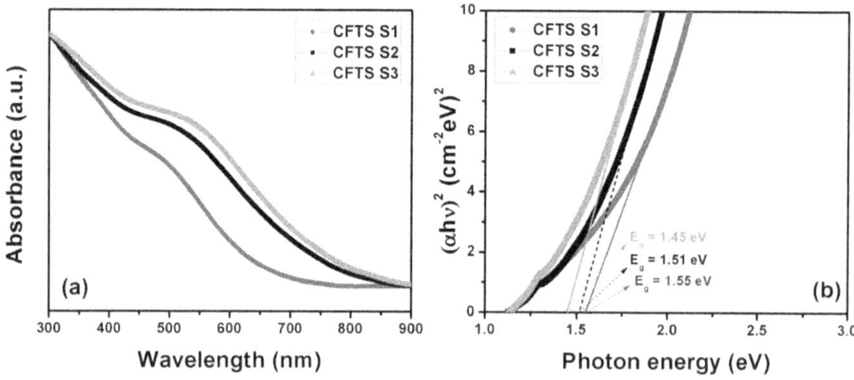

Figure 2.19 (a) UV-Vis spectra and **(b)** corresponding plots of hv vs (ahv)^2 to determine the optical band gap of sulfurized CFTS S1, CFTS S2 and CFTS S3 nanoparticles.

The band gaps gained from the Tauc plot for the sulfurized samples indicates that the band gaps have increased from 1.39 to 1.45 eV, 1.41 to 1.51 eV, and 1.42 to 1.55 eV for CFTS S1, CFTS S2, and CFTS S3 respectively from the as-synthesized sample as can be obtained as from **Fig. 2.19 (b)**. The change in the band gaps was due to the presence of surface defects on the CFTS nanoparticles. Owing to these defects, the absorption edge in the conduction band is shifted to the higher energy states. As a result, the energy states near the conduction band will be crowded and thus the optical band gap has increased [106].

(d) *I-V* characteristics of CFTS nanoparticles

We also investigated the photoresponse properties of the sulfurized samples by developing drop casting films of sulfurized CFTS nanocrystals in between two lateral Al electrodes on glass. The sulfurized samples also exhibits similar behaviour like the as synthesised sample with CFTS S2 nanoparticles as most photoresponse among them as shown in **Fig. 2.20 (a-c)**.

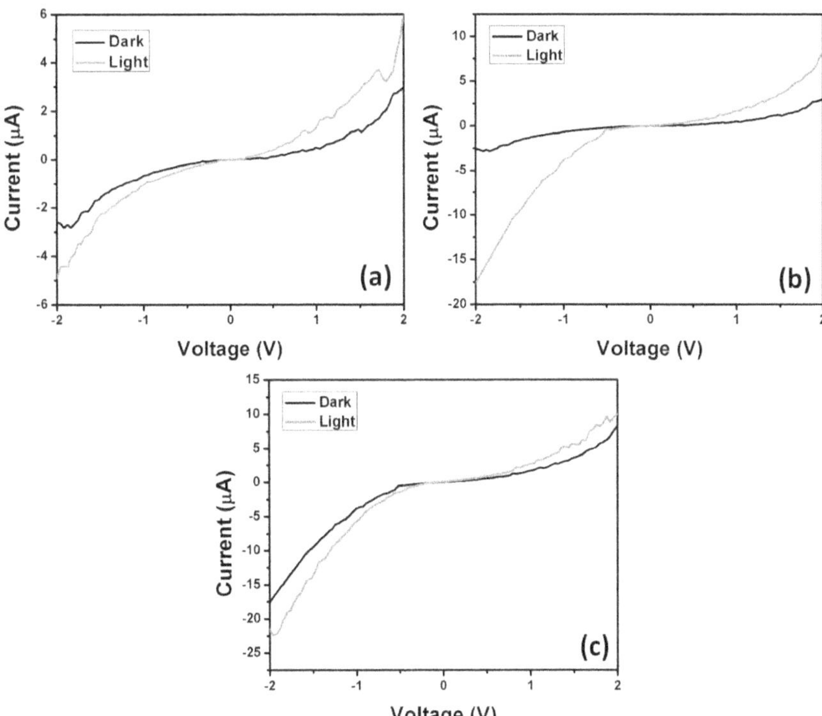

Figure 2.20 *I-V* characteristics of **(a)** CFTS S1, **(b)** CFTS S2 and **(c)** CFTS S3 nanoparticles in dark and under light illumination.

(e) Cyclic voltammetry (CV)

Fig. 2.21 illustrates the CV curves of sulfurized CFTS S2 nanoparticle at a scan rate of 50 mV/s and remaining scan rates are shown below. Here also the curves were measured in a potential ranging from -0.1 to 0.5 V with a scanning speed of 5, 20, 50, 80 and 100 mV/s, under dark and light illuminated conditions as shown in **Fig. 2.22 (a)** and **Fig. 2.22 (b)** for the CFTS S2 nanoparticles. At lower scan rates, the sulfurized samples also showed a similar symmetry with good electrical double layer capacitor (EDLC) characteristics. From the comparison of both, for the area under the curve for the dark and light condition, light illuminated sample showed larger capacitance when compared to the dark condition for both the as-prepared CFTS nanoparticles.

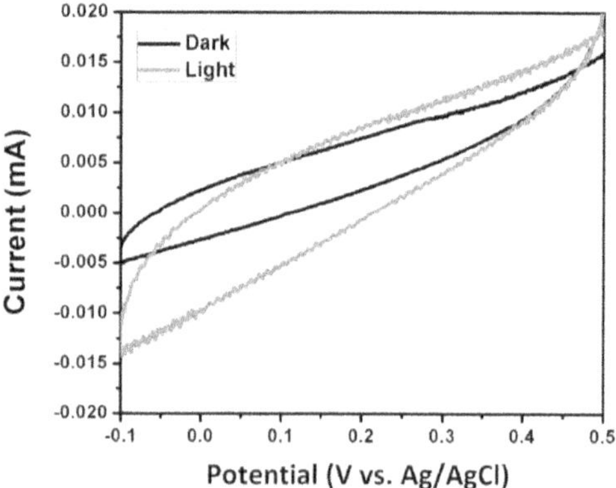

Figure 2.21 CV curves of sulfurized CFTS S2 nanoparticles at 50 mV/s scan rate in dark and under light illumination.

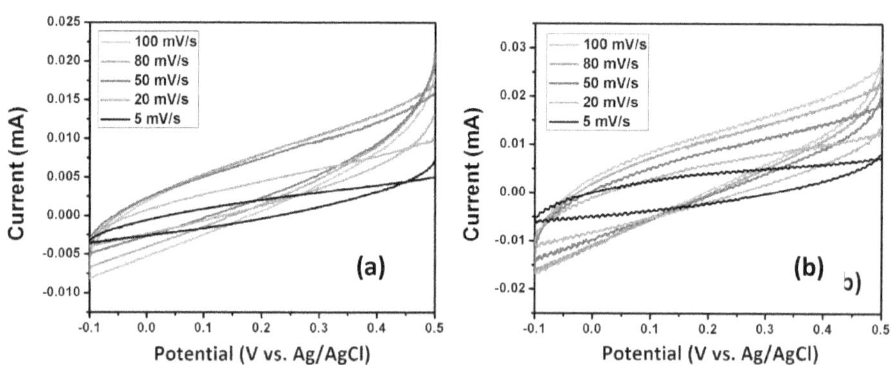

Figure 2.22 CV curves of CFTS S2 nanoparticles at various scanning rates (5, 20, 50, 80, 100 mV/s) in **(a)** dark and **(b)** under light illumination.

Fig. 2.23 summarised the curves of specific capacitances with varying scan rate of the sulfurized CFTS2 nanoparticles. In the sulfurized CFTS nanoparticles also, on increasing the scan rate the capacitance of the electrode is reduced. These results indicate that, in the sulfurized samples provide greater advantages in electrode-electrolyte interface along with large interface in the generation of photon excited electrons and holes. Similar studies were done for the sulfurized CFTS S2 samples, but the capacitance is reduced when compared with the as-prepared CFTS nanoparticles probably because of the presence of the secondary phases in the sulfurized samples.

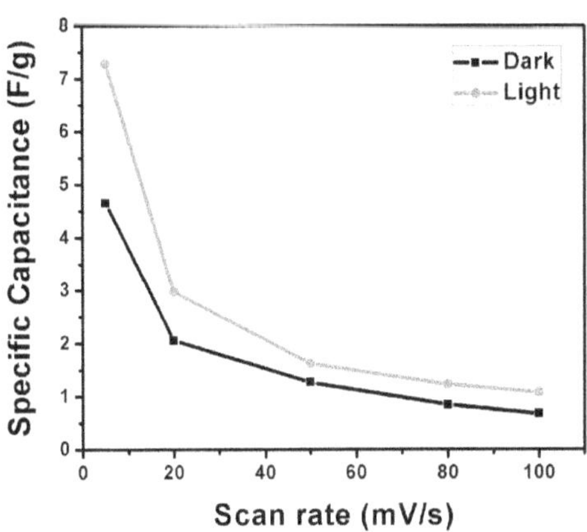

Figure 2.23 Comparative plot of specific capacitance as a function of scan rates.

(f) Galvanostatic charge/discharge (GCD)

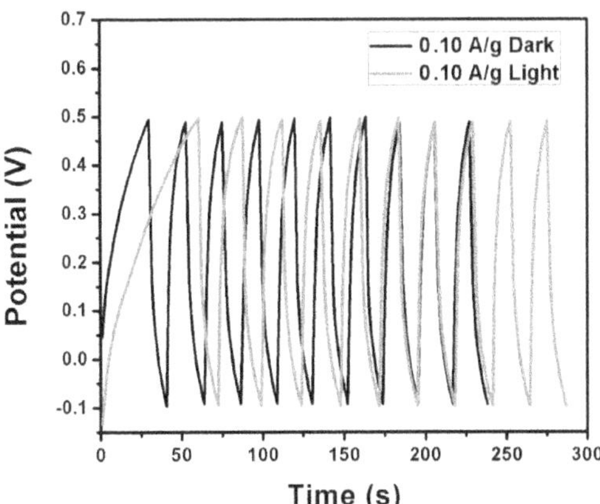

Figure 2.24 Galvanostatic charge/discharge profiles of sulfurized CFTS S2 nanoparticles at a constant current density of 0.10 A/g in dark and under light illumination respectively.

The GCD of sulfurized CFTS nanoparticles were also studied at a constant current density of 0.10 A/g in similar condition and the results are shown in **Fig. 2.24**. GCD measurement also reveals that the CFTS S2 is also exhibiting higher photo-enhanced super capacitive behaviour. The increase in the charge/discharge time suggests the enhancement of capacitance in the light irradiation.

2.4. Conclusion

CFTS nanoparticles were successfully synthesized with the help of a solvothermal route using PEG as a solvent followed by sulfurization. The crystallinity of the CFTS nanoparticles drastically increased after sulfurization which was confirmed by XRD and Raman spectroscopy. The bandgap of the as-prepared CFTS nanoparticles was found to be increased with the increase in the chain length of the solvents (1.39 to 1.42 eV). Whereas, the bandgap of sulfurized CFTS nanoparticles decreased with increase in the chain length of PEG. Few secondary peaks were observed after sulfurization, which leads to a decrease in the performance of the CFTS nanoparticles. Among the as-prepared samples, CFTS2 showed enhanced photoresponse behaviour than its counterparts (CFTS1 and CFTS3). Interestingly, the CFTS2 sample shows an excellent light sensitivity than the CFTS S2 nanoparticles, where it excites the charge carriers with help of light illumination and significantly helps in increasing the specific capacitance. It is believed that the excess sulfur on the surface plays a vital role in photo-enhanced defect-activated charge storage in CFTS. It is known that quaternary chalcogenide CFTS is a potential candidate for the photovoltaic absorber layer and also in the future due to its photo enhanced supercapacitive behaviour, this can also be used in supercapacitor applications.

Chapter 3
Influence of stabilizers and the synthesis of photoactive Cu$_2$FeSnS$_4$ thin films via sol-gel method

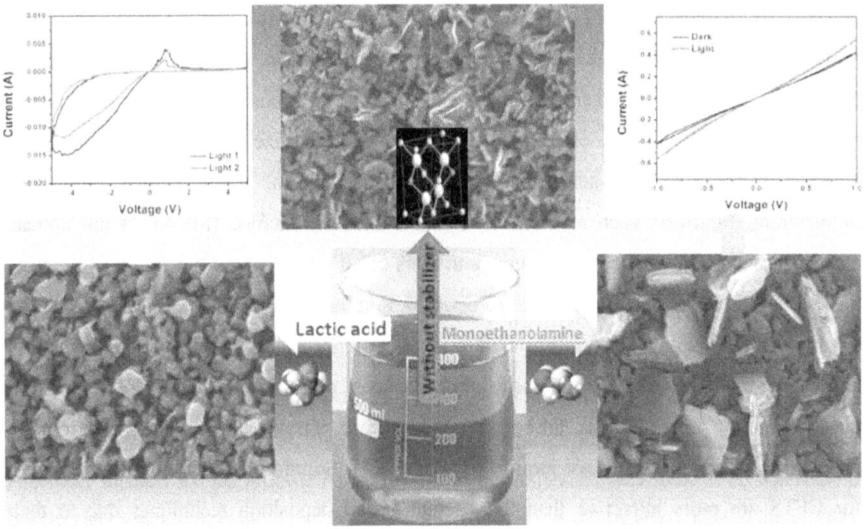

3.2.2 CFTS thin film fabrication by spin coating

In a typical spin coating technique, the active layer CFTS was coated from a single precursor. The cationic precursor solutions were synthesized in separate beakers and finally, all of the cationic precursor solutions consisting of $CuCl_2$, $Fe(NO_3)_3.9H_2O$, and $SnCl_2$ in 3 mL 2-methoxy ethanol each, in a stoichiometric molar ratio of 0.38:0.25:0.2 were mixed and stirred for one hour. Then a separate beaker containing 1.6 M $SC(NH_2)_2$ in 3 mL of 2-methoxy ethanol was used as the anionic precursor solution. Initially, $Fe(NO_3)_3$ solution was added into the $SnCl_2$ solution which then resulted in a golden yellow solution, followed by the addition of $CuCl_2$ solution exhibited greenish yellow coloured solution and finally after the addition of 1.6 M $SC(NH_2)_2$, the solution turned to light green colour.

Then the active absorber sol solution was spin-coated via spin coating route onto an ITO substrate at the rate of 2500 rpm for 30 s followed by post heating after deposition the substrate at 250 °C for 2 min on a hot plate. The coating was repeated twelve times to obtain a thick CFTS absorber thin film. The thin films were then sulfurized to form CFTS thin films by annealing under sulfur at 500 °C for 30 min followed by cooling down naturally to room temperature. The heating rate was set at 5 °C/min.

3.2.3 CFTS thin-film bistable device fabrication

At first, the indium tin oxide (ITO) glass substrates were cleaned with a soap solution (Dettol liquid hand wash), followed by deionized water, ethanol, acetone, and isopropyl alcohol (IPA) each for 10 minutes sonication. The ITO was etched on one side and then the active layer CFTS was deposited by spin coating technique. Finally, the aluminium was deposited by thermal evaporation method which led to the formation of the top electrode which is orthogonal to the ITOs. Whereas in the case of sulfurized films the top electrodes were fabricated by applying the silver paste on top of the film and the photoresponse was measured.

From here onwards the as-prepared CFTS thin films will be labelled as as-prepared CFTS-WS (CFTS fabricated without any stabilizer), as-prepared CFTS-L (CFTS fabricated with lactic acid) and as-prepared CFTS-M (CFTS fabricated with monoethanolamine) thin films and the sulfurized CFTS thin films will be labelled as sulfurized CFTS-WS, sulfurized CFTS-L, and sulfurized CFTS-M thin films respectively.

3.3 Results and Discussion

3.3.1 As-prepared CFTS thin films

(a) X-ray diffraction

The typical XRD patterns of as-prepared CFTS-WS, CFTS-L and CFTS-M are shown in **Fig. 3.1**. The diffraction peaks are less intense and broad which cannot be indexed. The main intense peaks for the as-prepared CFTS samples are not present which confirms that the formation of the stannite phase has not been completed [141]. Hence, a high temperature annealing in the presence of sulphur known as sulfurization is performed for these as-prepared CFTS thin films to make it crystalline in nature.

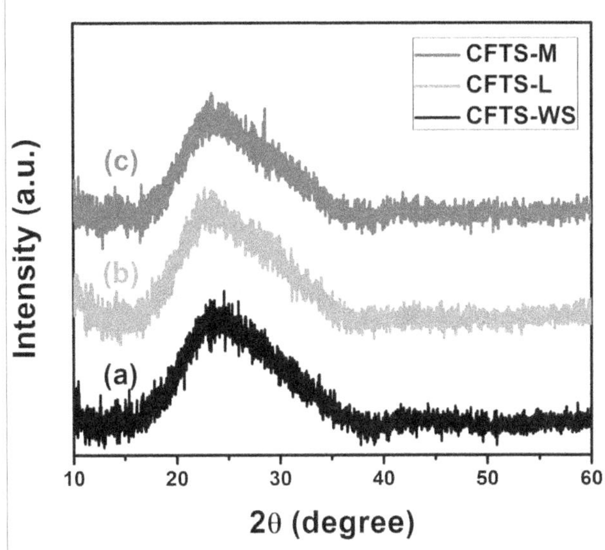

Figure 3.1 XRD patterns of as-prepared (a) CFTS-WS, (b) CFTS-L and (c) CFTS-M thin films.

(b) Raman spectroscopy

Raman spectroscopy is the complementary tool to confirm structural details identified by XRD. These problems make the XRD interpretation of CFTS thin films a difficult task and also may result in the wrong identification of the peaks if they are formed due to secondary phases. Hence, Raman spectroscopy was performed to check the presence of these secondary phases. Raman spectra of the as-prepared CFTS thin films are shown in **Fig. 3.2**.

The as-prepared CFTS thin films did not result in any considerable intense peaks. From Fig. 3.1, XRD reveals amorphous nature of the films. That could be the reason for not observing any Raman peaks.

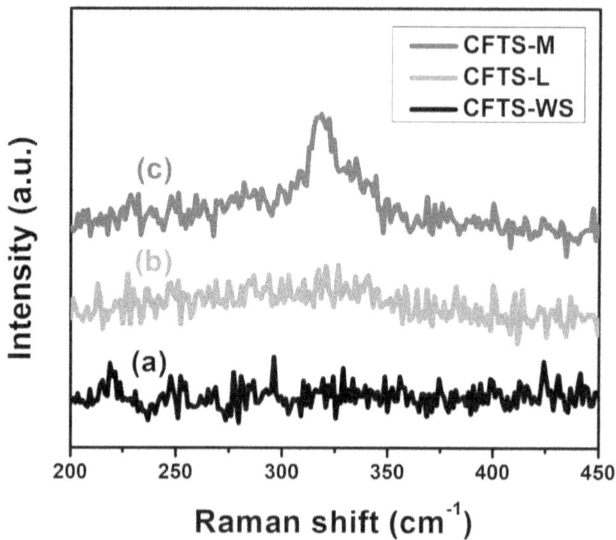

Figure 3.2 Raman spectra of as-prepared (a) CFTS-WS, (b) CFTS-L and (c) CFTS-M thin films.

(c) Field emission scanning electron microscopy (FE-SEM)

In addition to the structural, composition, morphological analysis is also crucial to evaluate the particle size and applicability of the films. **Fig. 3.3** shows the field-emission scanning electron microscopy (FE-SEM) images depicting the surface morphology of the as-prepared and sulfurized CFTS thin films. The FESEM images of as-prepared films show that all the films have a compact and uniform arrangement of smaller grains. Larger grain size and minimal defects are the properties that are required to improve the performance of the solar cells.

Also, SEM reveals that the sizes of the nanoparticles vary and are in a range of 15-20 nm and are arranged uniformly in the case of CFTS-L **(Fig. 3.3 (b))**. Whereas, for CFTS-M the arrangement is not uniform with small particle agglomeration occurs on top of the smooth particle surface **(Fig. 3.3 (c))**. Due to which entrapment of charge carriers occurs, as a result, it exhibits electrical bistability for both the CFTS thin films synthesized with the help of stabilizers. Whereas, the CFTS exhibits lesser defects and comparatively larger particles **(Fig. 3.3 (a))**. As it is evident from the SEM images, the particles are uniformly arranged on top of the smoothly arranged particle surface, which results in the free flow of the charge carriers. CFTS-L and CFTS-M show smaller grain size which is not desirable for high-efficiency solar cells.

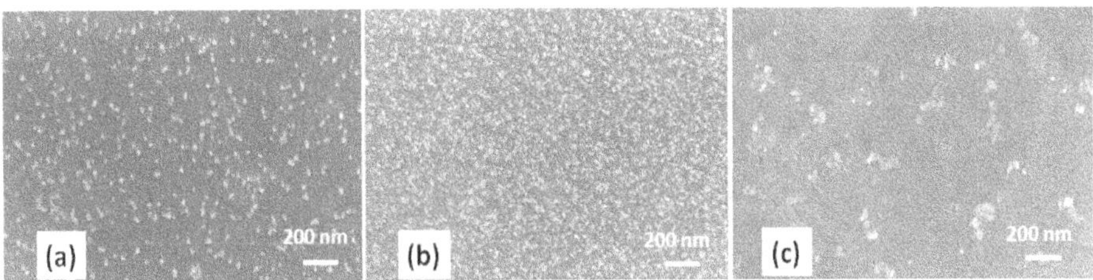

Figure 3.3 SEM images of the as-prepared **(a)** CFTS, **(b)** CFTS-L and **(c)** CFTS-M thin films.

(d) Energy dispersive X-ray spectroscopy (EDS)

The elemental composition of the as-prepared samples was determined using energy-dispersive X-ray spectroscopy (EDS) analysis as shown in **Fig. 3.4**. From **Fig. 3.4**, it can be seen that the elemental composition lacks a deficiency of stoichiometric ratio because of not forming the appropriate structure or due to incomplete formation of the required phase. As the Fe content is more and Cu content is less than the required stoichiometry. This specifies that the films are Fe-rich and Cu-poor. The ideal stoichiometric molar ratio should be 2:1:1:4. Since, the atomic ratios obtained in this study are not following the required molar ratio (2:1:1:4), thus the stoichiometry is not achieved for the as-prepared samples.

The losses due to the component diffusion into the substrates or the losses occurring due to the volatility of the components are the reasons for the stoichiometric variations due to the deposition methods. From these results, it can be seen that there is an obvious change in the elemental composition of the CFTS thin films when these thin films were fabricated with the help of stabilizers (CFTS-L and CFTS-M). Hence, the sulfurization of the as-prepared CFTS thin films was carried out, so as to form highly crystalline thin films with required stoichiometry.

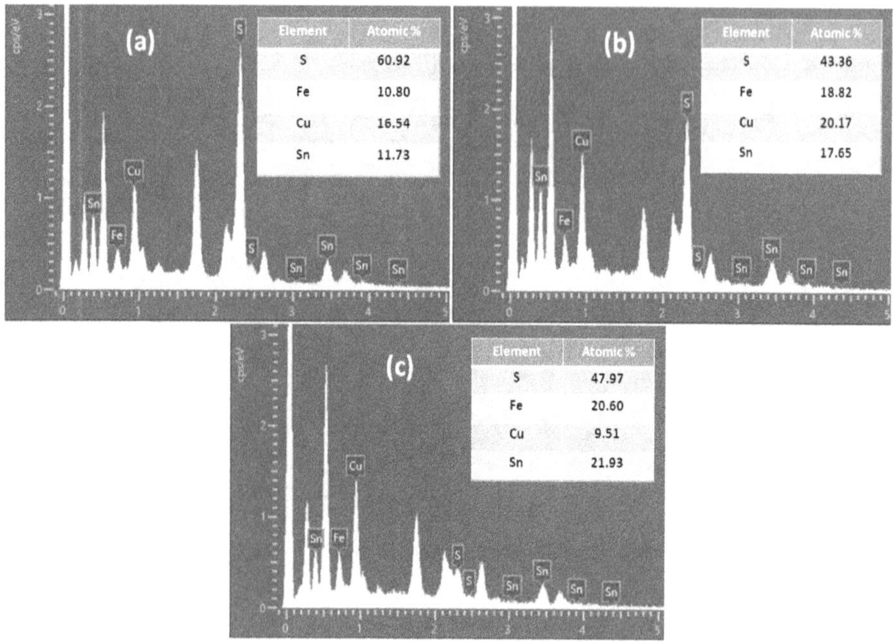

Figure 3.4 EDX spectra of as-prepared **(a)** CFTS, **(b)** CFTS-L and **(c)** CFTS-M thin films.

(e) UV-Vis spectroscopy

Since CFTS is a suitable candidate for solar energy harvesting devices, the estimation of the band gap is crucial, and this can be done by extrapolating the linear region of $(\alpha h\upsilon)^2$ vs $h\upsilon$ plots and E_g is derived from the intercept of the $h\upsilon$ axis. The as-prepared CFTS thin films were having a higher band gap when compared with the sulfurized CFTS thin films prepared by adding different stabilizers, probably due to the presence of defects in the films. Similarly for the as-prepared CFTS thin films also exhibit similar kind of absorbance and reflectance.

(f) *I-V* characteristics loops of CFTS thin films

Fig. 3.5 shows the electrical measurement results. **Fig. 3.5 (a-c)** depicts the current-voltage (*I-V*) characteristics of the as-prepared CFTS bistable device in a complete cycle between -5.0 V and +5.0 V. The coexistence of resistive switching and negative differential resistance (NDR) properties with excellent repeatability is exhibited. It shows that the value of the current is higher when voltage is swept from negative (-ve) to positive (+ve) values.

Also, we know that the current increase with the increase in the voltage as the current is proportional to voltage. However, under a positive voltage polarity, it achieves a high resistance state (HRS) from a low resistance state (LRS) as a result of which the current gets dropped sharply, exhibiting NDR behaviour. And the current increases sharply and achieves LRS from HRS due to the formation of conductive filaments. Thus, the change from HRS to LRS is due to the formation of the conducting filaments. When the voltage is reversed from +5.0 to -5.0 V, a drop in the current value was observed and the current was very small and almost unchanged and then increased with the increase of the voltage. These changes are observed because of the memristive mechanism which arises due to filamentary or interface type resistive switching.

As the former resistive switching occurs because of the inhomogeneity of the thin films and originates due to the formation of conductive filament known as a filamentary type of resistive switching. And the latter type is known as interface type resistive switching occurs due to the formation of Schottky contact resistance between an insulator and the electrode. Also, charge trapping/de-trapping happens when a forward bias is applied and thus the electrons can be drawn out from the traps, as a result, many positively charged carriers are left behind. Thereby, reducing the built-in potential by providing extra potential and thus the device is switched to LRS.

And when a reverse bias is applied to the device, the injected electrons can be trapped by these empty traps. Consequently, the device gets switched back to HRS. This was observed for CFTS-L and CFTS-M as shown in **Fig. 3.5 (b)** and **Fig. 3.5 (c)** respectively. Among them, CFTS-L was exhibiting enhanced electrical bistable behaviour, as the noise level in the plots was very much less compared to CFTS-M. Whereas for CFTS alone there was not much charge trapping happening, as a result, the electrical bistability is diminished as shown in **Fig. 3.5 (a)**. Furthermore, consecutive loops for all the three CFTS samples were analysed to study its stability and repeatability. All the CFTS samples were having a constant change of current with the applied voltage in dark and under the light condition as shown in **Fig. 3.6, Fig. 3.7,** and **Fig. 3.8**.

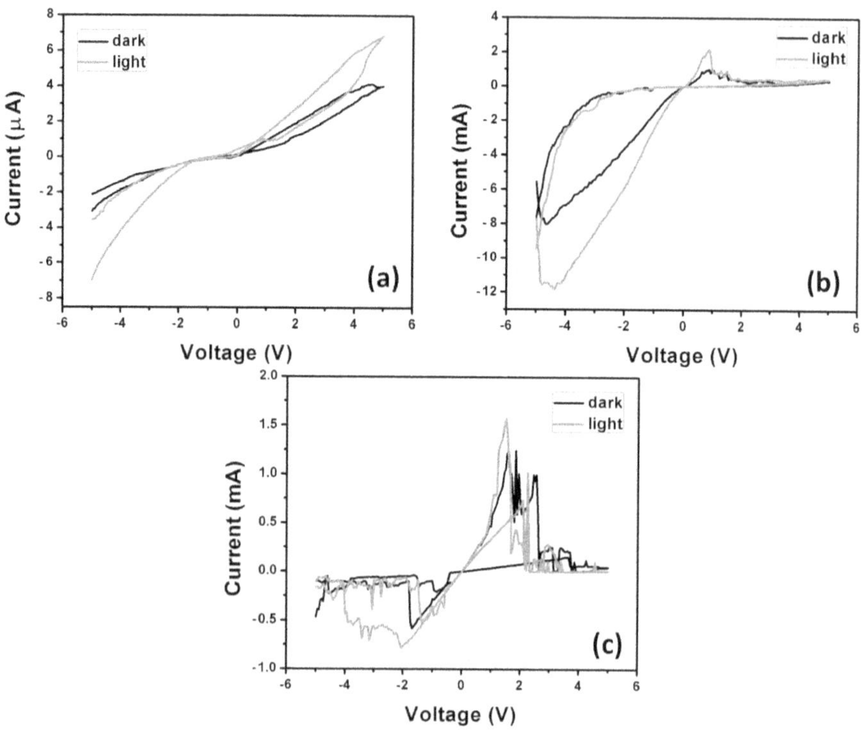

Figure 3.5 *I-V* characteristics in dark and under light illumination of as-prepared **(a)** CFTS-WS, **(b)** CFTS-L, **(c)** CFTS-M thin films.

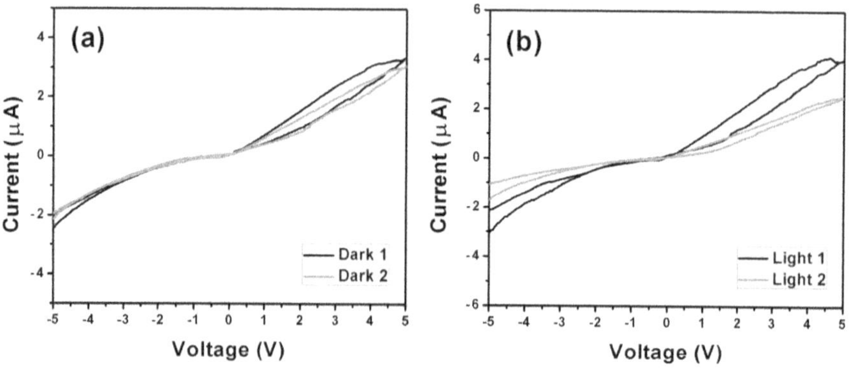

Figure 3.6 *I-V* characteristics showing consecutive loops in **(a)** dark and **(b)** under light illumination of as-prepared CFTS thin films.

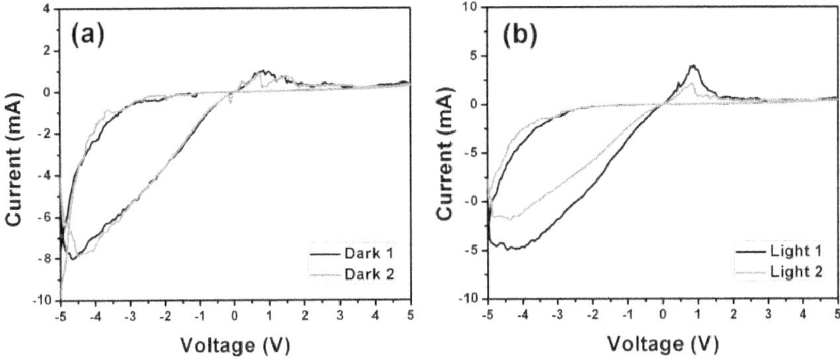

Figure 3.7 *I-V* characteristics showing consecutive loops in (a) dark and (b) under light illumination of as-prepared CFTS-L thin films.

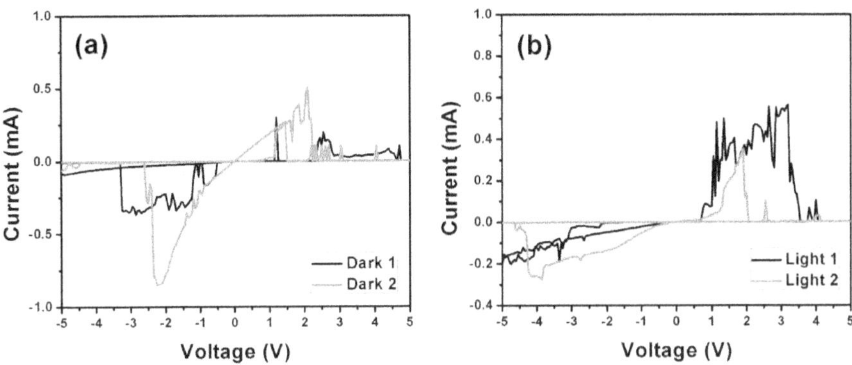

Figure 3.8 *I-V* characteristics showing consecutive loops in **(a)** dark and **(b)** under light illumination of as-prepared CFTS-M thin films.

(g) Write-read-erase-read" sequence

Since the I-V attributes are appearing on and off states, we have studied the "write-read-erase-read" sequence for various applications like random access memory (RAM). A pulse voltage to "write" or print a high conducting state and "erase" it, -1.5 *V* and 5.0 V biases were applied respectively. The current was measured under 1.0 V for either the states to be "read" or being probed for all the three different CFTS devices. The lower panel for **Fig. 3.9 (a), (b),** and **(c)** shows the current values depending on the write or erase values under the probe voltage. The RAM applications of these devices are shown by the change in the difference in the current values during the high and low conducting states. It also shows that the device can behave like a flip-flop between the two states. When an appropriate negative bias is applied, the charge carrier gets trapped in the grain boundaries and the interlaying layers. The charge carriers that are trapped act as steps for other charge carriers in enhancing the conductivity between the electrodes and the conductivity will be reduced. And by applying a suitable positive bias current, the trapped charged carriers will become free and get removed due to which the conductivity will be increased.

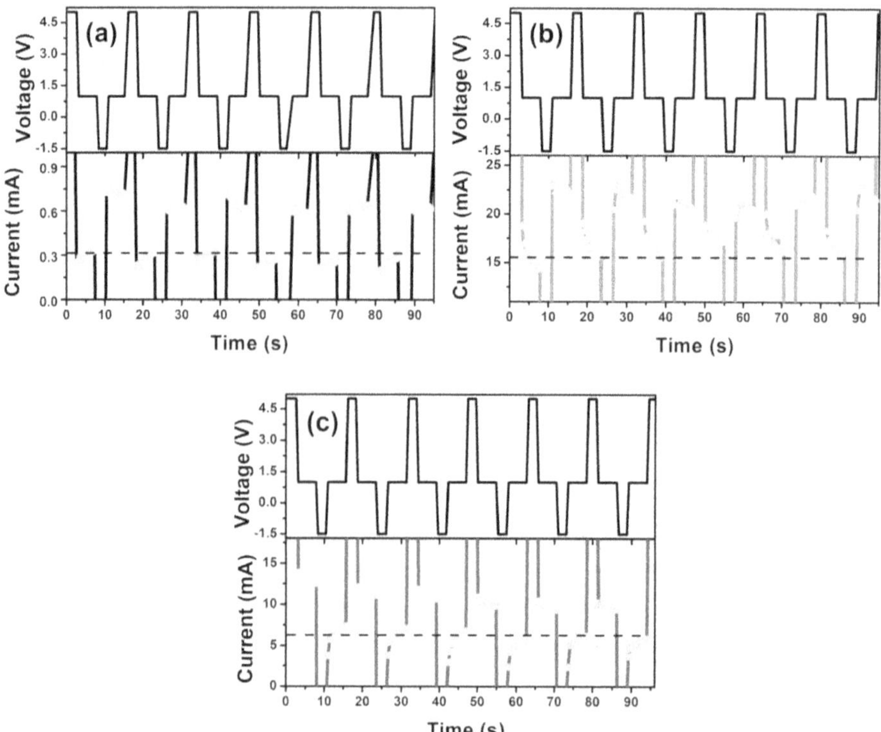

Figure 3.9 "Write-read-erase-read" sequence for random-access memory (RAM) applications. The voltage sequence and the corresponding values of device current as a function of time are shown in the upper and lower panels respectively. Amplitude of voltage pulse were +5 V and -1.5 V to "erase" and "write", respectively. Between the "write" and "erase" pulses, the high and low states were "read" by applying -1.0 V by measuring the current every 0.5 s.

3.3.2   Sulfurized CFTS thin films

3.3.2.1 Characterization of sulfurized CFTS nanoparticles

(a) X-ray diffraction

The typical XRD peak patterns of sulfurized CFTS thin films using different stabilizers such as monoethanolamine (MEA) and lactic acid are shown in **Fig. 3.10**. The highly crystalline phase of the sulfurized films characterized by the XRD is shown in **Fig. 3.10**. The diffraction peaks are of high intensity with less FWHM, indicating the crystalline nature of the sulfurized CFTS thin films. All the intense peaks in the XRD patterns are indexed with the JCPDS card no. 74-1025 can be assigned to the stannite phase with a tetragonal structure. The synthesized thin films show the main characteristic peaks at 23.0 °C, 28.4 °C, 32.7 °C, 36.9 °C, 40.9 °C, 44.9 °C, 47.4 °C, 50.1 °C, and 56.7 °C which are assigned to (110), (112), (200), (202), (114), (213), (204), (222) and (303) reflections from the planes respectively. A small secondary phase depicted as "*" at 26.4 °C indicates the Sn_2S_3 phase [142]. Other than that no other trace of additional peaks has been observed in the patterns for all the three CFTS thin films.

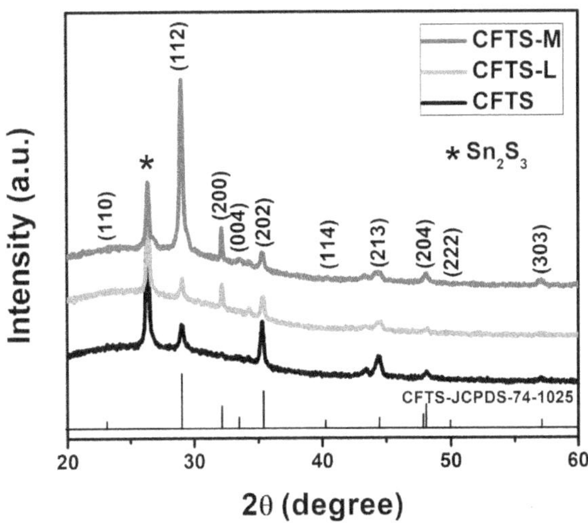

Figure 3.10 XRD patterns of sulfurized CFTS-WS, CFTS-L and CFTS-M thin films.

The crystallite size of CFTS, CFTS-L, and CFTS-M prepared using a sol-gel solution are calculated to be 28.45 nm, 28.04 nm, and 33.98 nm respective as listed in **Table 3.1**. It is very clear from **Table 3.1** that CFTS-M is larger compared to CFTS and CFTS-L. For the better performance of the solar cells, the films with larger grain sizes must be proficient for producing high-efficiency thin-film devices. The dislocation density (δ) and the strain (ε) of the films were calculated from the following equations (2) and (3),

$$\delta = 1/D^2_{hkl} \tag{2}$$

$$\varepsilon = \beta\, cos\theta/4 \tag{3}$$

The calculated dislocation density and microstrain for the CFTS-WS, CFTS-L, and CFTS-M thin films prepared using the sol-gel solutions are also shown in **Table 3.1**. The δ and ε values are larger for the CFTS-L thin films compared to the CFTS-WS and CFTS-M thin films. Hence after adding lactic acid as a stabilizer, the crystallite size decreases but the dislocation density and micro strain increase. Whereas, in the case of monoethanolamine the crystallite size increases while the dislocation density and micro strain decrease.

Table 3.1

Relevant parameters derived from the XRD patterns of the sulfurized CFTS-WS, CFTS-L and CFTS-M thin films prepared using sol-gel solutions.

Stabilizers	2θ (deg.)	FWHM of (112) peak (deg)	Crystallite size D (nm)	Dislocation density δ (lines per m^2)	Microstrain $\varepsilon = \beta/tan\theta$
Sulfurized CFTS-WS	28.5	0.288	28.45	1.23488E+15	1.11
Sulfurized CFTS-L	28.5	0.292	28.04	1.27172E+15	1.12
Sulfurized CFTS-M	28.5	0.241	33.98	8.6567E+14	0.93

(b) Raman spectroscopy

Raman spectroscopy was performed with a 532 nm excitation Laser source to study the phase purity of the material and to study the co-existing secondary phases that cannot be identified in XRD analysis. The Raman peaks of sulfurized samples were reliable with the XRD results and the peaks were observed at 281 cm^{-1} and 317 cm^{-1} (**Fig. 3.11**). These peaks correspond to stannite CFTS which is similar to that described in the literature. The major peak at 317 cm^{-1} corresponds to the asymmetrical vibration of a pure anion mode of a sulfur atom around the tin and the peak at 281 cm^{-1} is due to the pure anion mode around the copper cation [143][144].

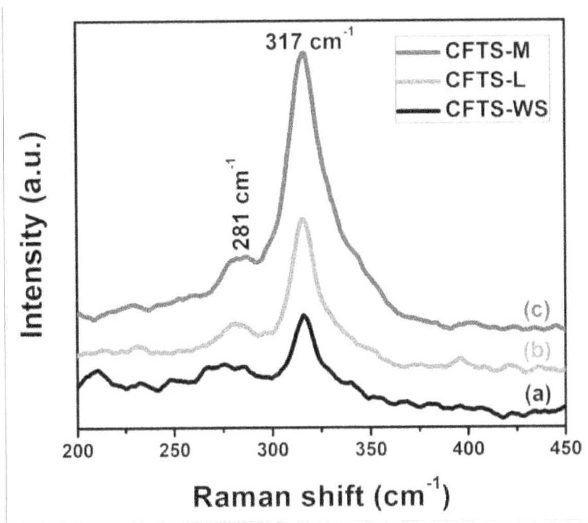

Figure 3.11 Raman spectra of sulfurized CFTS-WS, CFTS-L and CFTS-M thin films.

(c) Field emission scanning electron microscopy (FE-SEM)

All the as-deposited films are annealed at a higher temperature in the presence of sulfur vapour to achieve larger grain size and crystallinity and to eliminate the defects. So after sulfurization, it is seen that the grain sizes have increased with crystalline texture for all three samples due to the incorporation of sulfur into CFTS films **(Fig. 3.12 (a-c))**. Fewer grain boundaries are a result of larger grain boundaries which further causes lesser scattering of charge carriers. Thereby reducing the photo-generated charge carrier recombination at the grain boundaries and enhances the photoresponse. **Fig. 3.12 (a)** shows the growth of few thin flakes having 10 nm thicknesses on the surface of the uniformly arranged CFTS nanoparticle films. Whereas, CFTS-M consists of larger flakes that are grown on the surface, having a thickness of 50 nm without any noticeable voids as shown in **Fig. 3.12(c)**. And, CFTS-L shows larger grains and compact structures of thin films as shown in **Fig. 3.12(b)**.

Figure 3.12 FE-SEM images of sulfurized **(a)** CFTS-WS, **(b)** CFTS-L and **(c)** CFTS-M thin films.

(d) Energy dispersive X-ray spectroscopy (EDS)

The elemental composition of sulfurized samples was established using energy dispersive X-ray spectroscopy (EDS) analysis as shown in **Table 3.2.** However, from **Table 3.2** it very well may be seen that there is an elemental composition of sulfurized CFTS thin films after the addition of stabilizers. Whereas, the calculated atomic percentage ratios of Cu/A, Fe/A, and S/B showed the elemental composition of the respective samples, were A= Fe+Sn and B=Cu+Fe+Sn, in which Cu/A≤1 and Fe/S≥0.5. To acquire highly crystalline films, the spin-coated thin films were sulfurized at a higher temperature in the presence of the Argon atmosphere as mentioned in the experimental part. Notwithstanding, the cation content ratios of Cu/A, Fe/A, and S/B post-sulfurized samples changes, even though the overall stoichiometry was found to be admissible. And, these values prove that the samples were uniform throughout the scanned region of Cu, Fe, Sn and S are approximately 2:1:1:4, which is in the necessary stoichiometric proportion.

Table 3.2

EDS analysis of sulfurized CFTS-WS, CTS-L and CFTS-M thin films. Here A=Fe+Sn and B=Cu+Fe+Sn

Sample	Cu (at%)	Fe (at%)	Sn (at%)	S (at%)	Cu/A	Fe/A	S/B
Sulfurized CFTS-WS	19.30	15.63	11.20	53.87	0.71	0.58	1.16
Sulfurized CFTS-L	22.98	11.60	9.89	55.53	1.06	0.53	1.24
Sulfurized CFTS-M	22.46	11.63	9.60	56.31	1.05	0.54	1.28

(e) UV-Vis spectroscopy

The sulfurized CFTS thin films exhibit a strong and broad absorption peak in the UV and visible region as shown in **Fig. 3.13 (a)**. As can be seen clearly from **Fig. 3.13 (b)**, the optical band gap of sulfurized CFTS thin films was decreased after the addition of the stabilizers. The values of the band gaps are estimated at 1.48 eV for sulfurized CFTS-WS and 1.37 and 1.30 eV for CFTS-L and CFTS-M thin films fabricated with the help of lactic acid and MEA as stabilizers respectively which was well in the range of the essential ideal band gap for thin-film solar cells.

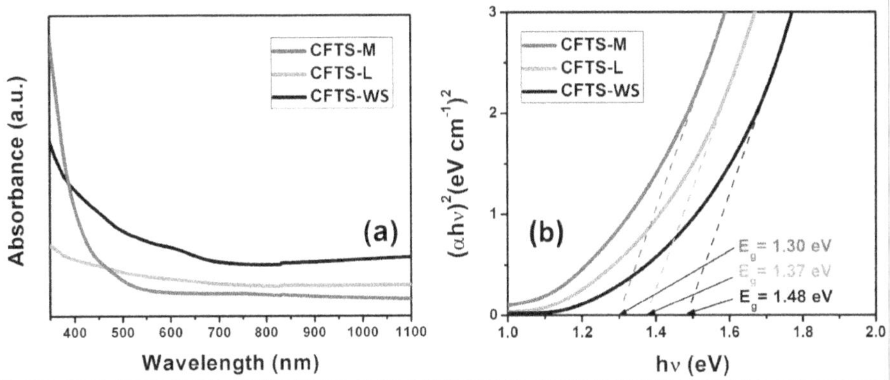

Figure 3.13 (a) UV-Vis absorption spectra and **(b)** Tauc plots of sulfurized CFTS-WS, CFTS-L and CFTS-M thin films.

(f) *I-V* characteristics of CFTS thin films

The *I-V* characteristics were tested for the sulfurized films, wherein it did not exhibit any electrical bistable behavior as shown in **Fig. 3.14**. This can be attributed to the elimination of the defects present in the films annealed at higher temperatures in the presence of sulfur. It is observed due to grain growth that occurs on the surface of the films which will reduce the grain boundary. This is succeeded by, the trapping of the charge carriers is eliminated and the free flow of the charge carriers occurs. The photoresponse of these films was also studied among them sulfurized CFTS-L was exhibiting a better response. The sulfurized CFTS-L and CFTS-M exhibit higher photoresponse as shown in **Fig. 3.14(b)** and **(c)** than the sulfurized CFTS thin films as shown in **Fig. 3.14 (a)**. Also, the current value has increased from 0.42 A for the sulfurized CFTS-WS to 0.57 A and 0.58 for CFTS-L and CFTS-M respectively when treated with various stabilizers (MEA and Lactic acid). The carrier mobility of the post-sulfurized CFTS thin films was increased. This further depicts that the sulfurized CFTS thin films fabricated with the help of stabilizers exhibit higher performance than the one fabricated without any stabilizer.

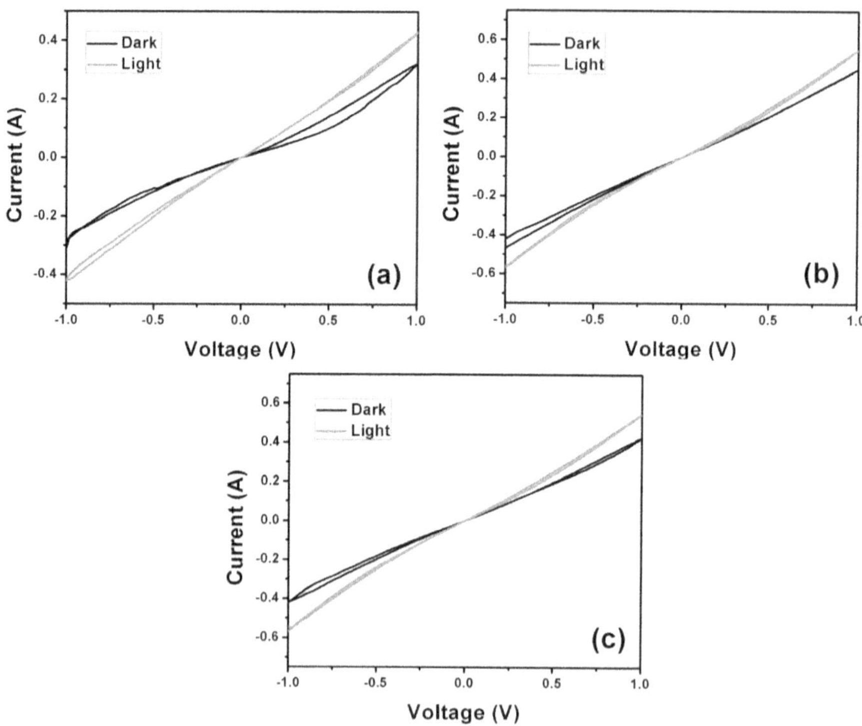

Figure 3.14 *I-V* characteristics in dark and under light illumination of sulfurized **(a)** CFTS-WS, **(b)** CFTS-L and **(c)** CFTS-M thin films.

3.4 Conclusions

In this report, three CFTS thin films were fabricated by the sol-gel method and spin-coating process using lactic acid and monoethanolamine as stabilizers and one without any stabilizers followed by sulfurization process. The X-ray diffraction patterns of as-prepared CFTS thin films did not show good crystallinity, whereas the sulfurized CFTS thin films exhibited highly crystalline peaks. The sulfurized CFTS thin films possess stannite structure and the band gaps of the CFTS, CFTS-L, and CFTS-M thin films are 1.48, 1.37, and 1.30 eV respectively. All the films exhibited electrical bistability before sulfurization among which CFTS fabricated with the help of stabilizers showed much better memory effect which can be depicted from read-write plots. However, the sulfurized CFTS thin films were devoid of electrical bistability behavior but showed good photocurrent responses and photo stability. It was due to the surface textures of the as-synthesized CFTS were not well aligned whereas the post-sulfurized CFTS thin films exhibited excellent crystalline nature with well grown grains. Thereby, the carrier mobility of the sulfurized samples was also increased, as a result the photocurrent also increased. Also, both as-prepared and sulfurized CFTS thin films exhibit suitable properties of implementation in the industrial process to fabricate thin-film solar cells.

Chapter 4

Electrical bistability and memory switching phenomenon in Cu$_2$FeSnS$_4$ thin films: role of p-n junction

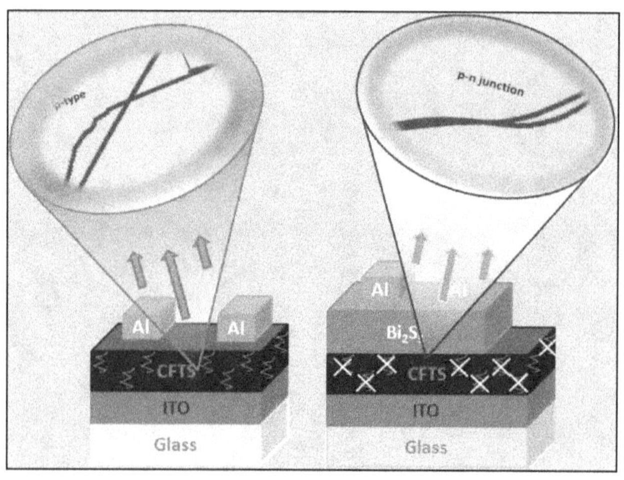

CHAPTER 4

Electrical bistability and memory switching phenomenon in Cu_2FeSnS_4 thin films and the role of p-n junction

4.1 Introduction

The CFTS thin films can be grown by several physical and chemical techniques such as hydrothermal and sol-gel as mentioned in the previous chapters. The major setback/bottleneck is the presence of secondary phases such as Cu_2S, SnS_2, ZnS during the growth of CFTS which was discussed in the previous chapters. Furthermore, the stability of CFTS nanoparticles was very weak due to the chemical potential of quaternary material which causes off-stoichiometric formation. In order teffectivly apply CFTS for commercial applications, the presence of secondary phases and other unwanted defects, that occur during the growth of CFTS, must be minimized or avoided.In this chapter, we have reported a technique to remove all the secondary phases in CFTS thin films known as the SILAR method. CFTS thin films were grown on different substrates to investigate the role of substrates. Based on the results ITO was found to be the best substrate for the growth of stoichiometric CFTS thin films.

SILAR is an exceptional film-fabrication technique employed to deposit films onlarge-area substrates. Recently, numerous works are reported on memory devices by the use of metal oxides and have promising potential applications towards future technologies. The copper-based quaternary chalcogenides have drawn the attention of research and development in the field of next-generation thin-film absorber materials for multifunctional applications. Quaternary photovoltaic materials have interested electrical and electronics researchers for their electrical, optical, and magnetic properties due to their high absorption coefficient, environmental stability, direct band-gap, and easy availability. In particular, Cu_2FeSnS_4 (CFTS) has properties comparable to that of CZTS, with a comparable band-gap energy (1.2-1.5 eV) [43]. CFTS can also be used in various optoelectronic applications as an alternative to CZTS since both CZTS and CFTS have analogous structures and properties [50, 145]. Hence, plausible application of CFTS can be found in memory applications [105], water splitting [146], and solar cells [48, 61]. Also, a study was carried out for a p-n junction (CFTS and Bi_2S_3) device to realize the effect of depletion potential and conductive filament formation in the electrical bistability mechanism.

A phase pure and highly crystalline stannite CFTS thin film is fabricated with the help of the SILAR method and its maiden study in the memristive application is explored. Memristive application or electrical bistability has been demonstrated by different materials[138, 147] such as CZTS [140], CdSe [137], ZnO [148, 149], and perovskites [150,151] etc. However, quaternary chalcogenides such as CFTS is not utilized for memory applications are not investigated much except for in a few studies [140]. Electrical bistability as hysteresis characteristics is observed in various materials such as metal chalcogenides synthesized via solution-based methods. In recent years, steep growth is seen in the use of semiconducting nanoparticles for memory applications. Memory devices are considered as one of the most capable devices for advanced electronic applications due to their low power consumption and simple device structure. Amongst these memory devices, the random-access memory (RAM) devices which are based on resistive switching are an integral part of current computers and other modern electronic devices [152, 153]. Mostly, resistive switching RAMs (RRAM) are used in memristive and neuromorphic systems [63, 153]. Redox based RAM (ReRAM) is a kind of RRAM which is more common. Recently, Valov et. al. discussed the physical phenomenon for ReRAM at the atomic level [154]. The existence of a memristor was initially predicted by L. Chau in 1971 [62].

Recently, Zhang et. al., with the aid of an in-situ transmission electron microscopy (TEM) demonstrated the formation of a resistive switching mechanism in a p-n junction [155]. Whereas, we observed that the formation of a depletion region prevents the formation of conductive filament which inhibits the resistive switching behavior in p-n junction devices. The study was performed with basic structures consisting of ITO/p-type CFTS/Al and ITO/p-type CFTS/n-type Bi_2S_3/Al. The current-voltage characteristics of the device showed two distinct paths in the low and in the high conducting states. This work also includes the characterization of the device fabricated with n-type (Bi_2S_3) alone and confirms that the device fabricated with only p-type CFTS shows the resistive switching behaviour. The detailed analysis discloses that the switching or the electrical bistability results from ion migration and refutes the conductive filament formation.

4.2 Experiment

4.2.1. Chemicals and materials required

The chemicals used are copper sulfate pentahydrate (99%, Thomas Baker), iron nitrate nonahydrate (99%, Thomas Baker), tin chloride dihydrate (99.9%, Thomas Baker), sodium sulfide (99%, Merck), deionized water (DI water), indium tin oxide (ITO, 8-12 Ω/sq, BAT-SOL, Australia), aluminium pellet (3-5 mm, 99.99%, Sigma-Aldrich, Germany). All these materials were used without further purification.

4.2.2 Preparation of precursor solution for pure CFTS thin films

The active layer CFTS was coated with the help of successive ionic layer adsorption and reaction (SILAR) method. The cationic precursor solutions were prepared in separate beakers reported elsewhere [6] and finally, all of the cationic precursor solutions consisting of $CuSO_4.5H_2O$, $Fe(NO_3)_3.9H_2O$ and $SnCl_2.2H_2O$ in 3 mL ethanol each, in a stoichiometric molar ratio of 2:1:1 were mixed and stirred for half an hour. Initially, $Fe(NO_3)_3$ solution was added to the $SnCl_2$ solution which resulted in a golden yellow colored solution, followed by the addition of $CuSO_4$ solution resulting in a greenish-yellow solution. A separate beaker with 4 M Na_2S in 9 mL ethanol was used as the anionic precursor solution. Then the active absorber material was coated via the SILAR route.

4.2.3 Preparation of precursor solution for Bi₂S₃ thin films

The n-type material (Bi_2S_3) was prepared by using 4 mM $Bi(NO_3)_3$ and 6 mM Na_2S in 15 mL DI water each in separate beakers and was coated on top of CFTS with the help of a similar SILAR route.

4.2.4 Thin film deposition and formation of CFTS and Bi₂S₃ thin films

The indium tin oxide (ITO) glass substrates were first cleaned with the help of ultrasonic treatment, initially by using solution (commercial hand wash), followed by deionized water, ethanol, acetone, and isopropyl alcohol (IPA) each for 10 minutes. Then the active layer CFTS was deposited with the help of the SILAR method. Initially, the ITO coated glass slide was dipped in the cationic precursor for 45 seconds, and then it was dipped in the solution containing anionic precursor for 60 seconds. After every dipping procedure, the substrate was rinsed with ethanol and air-dried. After each immersion procedure in the cationic and anionic solutions, a cycle of formation of the SILAR film was completed. The cycle was repeated 10 times to get a 300 nm thick CFTS thin film on the ITO glass substrate. All the depositions were carried out at room temperature and at the end of all depositions it was annealed at 150 °C for 30 minutes. Finally, the device fabrication was completed after the thermal deposition of aluminium under vacuum. Similarly, for the p-n junction device, a total of ten Bi_2S_3 layers were deposited to get a 150 nm thick Bi_2S_3 thin film on top of ten-layer CFTS thin film to complete a p-n junction device, and aluminum metal contact (300 nm of Al) was deposited on top by thermal evaporation method. The film thickness increases linearly up to ten cycles of Bi_2S_3.

4.2.5 Characterization of CFTS and Bi_2S_3 thin films

The structural and the phase purity of the fabricated CFTS and Bi_2S_3 thin films were studied by of X-ray diffractometer with the help of Ultima IV Model X-ray diffractometer (Rigaku) with Cu Kα irradiation. The morphological analysis of our CFTS thin films were characterized by a fieldtion.of NTEGRA Prima by NT-M microscope (FEcal analysis of our CFTS thin Atomic force microscope (AFM) was operated in the tapping mode with the help of NTEGRA Prima by NT-MDT (Russia). The current-voltage (I-V) characteristics of CFTS and the p-n junction (CFTS/Bi_2S_3) devices were measured using a Keithley 2450 source meter. Finally, the electrochemical impedance spectroscopy (EIS) of CFTS was recorded with the help of ZIVE SP1 Potentiostat/Galvanostat/EIS electrochemical workstation.

4.3. Results and Discussion

4.3.1 p-type CFTS thin film

4.3.1.1 Characterization of p-type CFTS thin film

(a) X-ray diffraction

The typical XRD pattern of the as-synthesized pure phase CFTS thin films is shown in **Fig. 4.1 (a)**. The observed diffraction peaks are in agreement with the stannite phase of CFTS with JCPDS Card No. 44-1476 with a tetragonal structure. The major diffraction peaks at 23.0°, 28.5°, 29.8°, 32.8°, 33.3°, 37.8°, 47.1°, 50.2°, 55.9°, 56.6°, 61.9° and 68.8° can be attributed to (110), (112), (103), (200), (004), (211), (220), (204), (222), (312), (116), (321) and (400) planes respectively [6]. In the previous chapters, we could see that the presence of impurities and secondary phases which could hinder the photo physical properties of the CFTS device. But in this method (SILAR) we were able to fabricate a phase pure CFTS thin film corresponding to tetragonal structure and space group of $1\overline{4}2m$ [47].

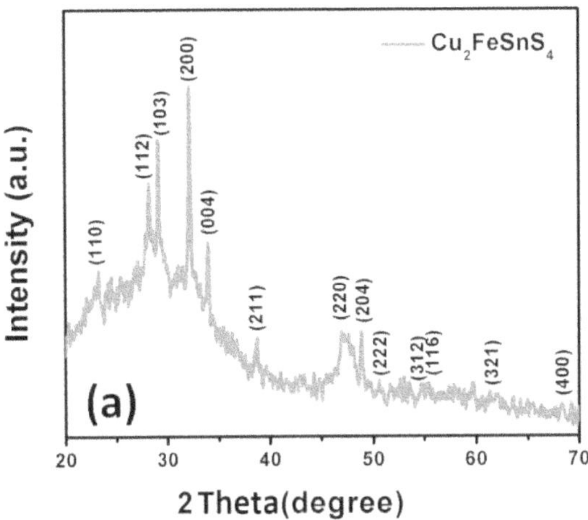

Figure 4.1 XRD pattern of CFTS thin films.

(b) Field emission scanning electron microscopy (FE-SEM)

Fig. 4.2 presents an FE-SEM image of the CFTS films. It can be observed that the surface is rough and has a large amount of non-uniform particles due to which the grain boundaries are different for different areas. As a result, the accumulation of charge carriers happens at one point in the grain boundaries and these accumulations restrict the flow of free charge transport [156]. Here, there is chance for the charge carriers to get trapped on the grain boundaries leading to the formation of copper conducting filaments which requires a minimum amount of threshold voltage to get rid of this trap and to take part in conduction.

Figure 4.2 The FE-SEM image showing the surface morphology of Cu$_2$FeSnS$_4$ (CFTS) thin film.

4.3.2 n-type CFTS thin film

4.3.2.1 Characterization of n-type Bi_2S_3 thin film

(a) X-ray diffraction (XRD)

For studying the electrical bistability of the p-n junction, we also synthesized the Bi_2S_3 thin films separately. It was found that the crystallinity and the phase purity of Bi_2S_3 thin films match quite well with the standard XRD diffraction pattern of Bi_2S_3 with JCPDS Card No. 17-0320 with an orthorhombic phase of Bi_2S_3 [37]. The peaks at 2θ values equal to 11.7°, 17.7°, 22.7°, 24.9°, 28.6°, 31.7°, 35.6°, 40.7°, 46.3°, 59.6°, and 60.9° can be indexed to (020), (120), (220), (130), (211), (221), (240), (141), (431), (242) and (171) planes of Bi_2S_3 respectively **(Fig. 4.3)**. The absence of secondary or impurity peaks confirms the purity of Bi_2S_3.

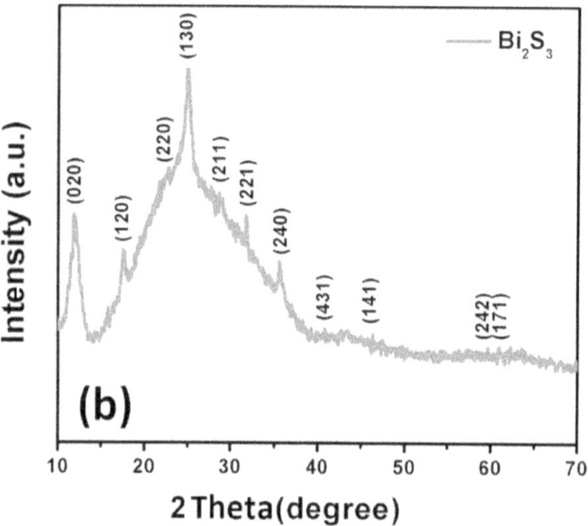

Figure 4.3 XRD pattern of Bi_2S_3 thin films.

4.3.3 Effect of p-n junction

The effect of adding a p-type layer results in the free electrons being able to cross the depletion region from one side to the other. The behavior of the p-n junction with regards to the potential barrier's width produces an asymmetrical conducting two-terminal device. A p-n junction device is one of the simplest semiconductor devices around, and which has the characteristic of passing current in only one direction only. It does not behave linearly with respect to the applied voltage as it has an exponential current-voltage (I-V) relationship and therefore we cannot describe its operation by simply using an equation such as Ohm's law. Here a p-type CFTS is combined with n-type Bi_2S_3 to form a p-n junction. SILAR films of several layers of p-type and n-type semiconductors have been considered so that both the components of p-n junction could be formed through the SILAR method and a top aluminium electrode to maintain the devices. Then the different characterizations were carried out to study the effect of a p-n junction in various memory aspects.

(c) Atomic force microscopy (AFM)

Fig. 4.4 (a) and **(b)** display the morphology of CFTS film as characterized by an atomic force microscope. The CFTS layer shows bigger lumps than the $CFTS/Bi_2S_3$ junction film. The AFM tip scans the surface of the film and does not probe deep inside to characterize the particle size. However, there are valleys on the surface the film looks boggy and bumpy **(Fig. 4.4 (a))**. Similar to what is observed here for the CFTS film. On the hind side, a completely covered surface would look much smoother. The case is similar to $CFTS/Bi_2S_3$ film **(Fig. 4.4(b))**, with the voids of the CFTS layers being filled with Bi_2S_3.

The films result to be more or less uniform and particles have been deposited isotropically. From the pictures, it is clear that the junction layer is more compact and the voids in the single-layer films have been covered with the next layers. Corresponding spreading resistance images are given in the Fig. 4.5 and it shows that the conducting species cover the substrate uniformly.

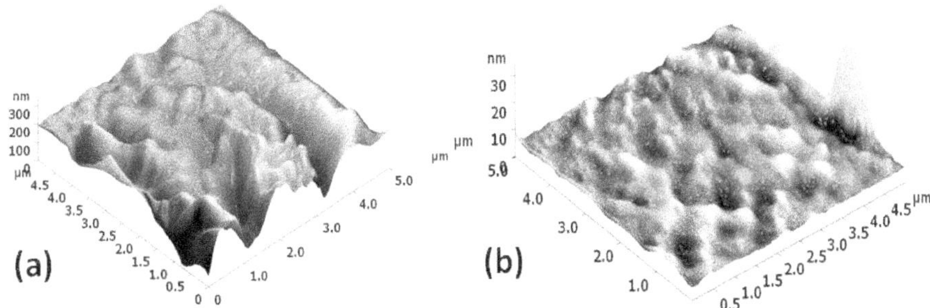

Figure 4.4 AFM images of **(a)** CFTS layers and **(b)** $CFTS/Bi_2S_3$ junction film.

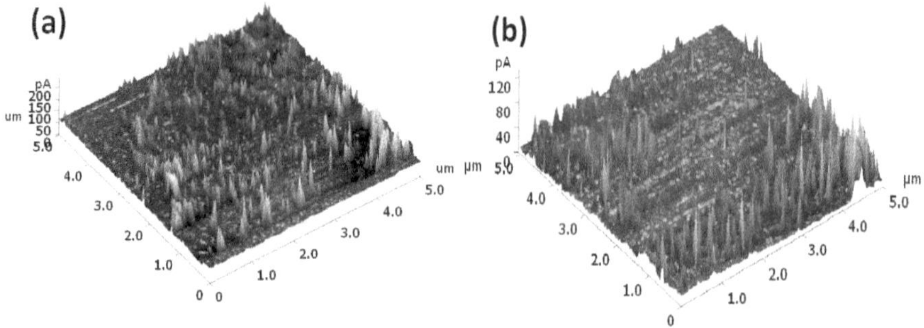

Figure 4.5 The spreading resistance images measured by the conducting AFM of **(a)** CFTS thin film and **(b)** CFTS/Bi$_2$S$_3$ p-n junction.

(d) *I-V* characteristic loop: Switching properties and mechanisms

Fig. 4.6 (a) and **(b)** shows the current-voltage (*I-V*) characteristics of the CFTS bistable device in a complete cycle between -1.0 V and +1.0 V. When voltage is swept from negative (-ve) to positive (+ve) values it is seen that the current is higher and also follows a straight-line path. The switching occurs at negative bias from a low conducting state to a highly conductive state. While sweeping, the device initially shows low conductivity (OFF-state) and when the applied voltage reaches a threshold value (-0.5 V), the device is switched to a highly conductive state (ON-state) as the current increases abruptly. The device then retains the ON-state during the subsequent sweeping process and thus shows the electrical bistability behaviour. This behavior can be attributed to the formation of conductive filaments and/or other arrangements of defect sites present in the film.

Similarly, for other bias voltage also this device shows the same trend, for example, 2 V as shown in **Fig. 4.7**. But on changing the voltage scan range threshold voltage range also changes, this is probably due to the energy of the defect sites will depend on the maximum voltage applied, so the threshold voltage will also change accordingly. The p-n junction (CFTS/Bi$_2$S$_3$) device does not exhibit any memory effect as the current-voltage loop does not show significant hysteresis behavior as shown in **Fig. 4.5 (b)**. This is attributed to the formation of a depletion region, preventing the formation of conductive filaments. So, no switching for p-n junction was observed.

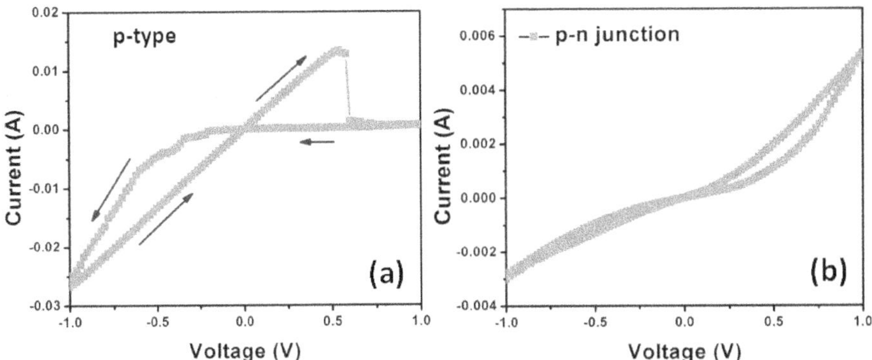

Figure 4.6 *I-V* characteristics of **(a)** p-type (CFTS) and **(b)** p-n junction (CFTS/Bi$_2$S$_3$) devices in a 1 *V* voltage loop.

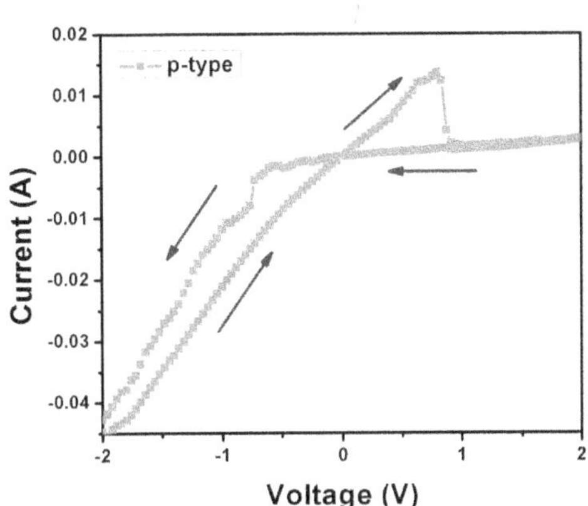

Figure 4.7 *I-V* characteristics of a p-type (CFTS) device swept at 2 *V* bias voltage loop.

In conducting AFM (C-AFM) the AFM tip probes the surface very closely, and consequently the current-voltage characteristics are essentially local, revealing the electronic nature of the surface morphology of the film. However, this may be influenced by the applied voltage during the C-AFM measurement, and the current is non-zero except for where the surface is highly conductive, which was verified in the current-voltage characteristics of the CFTS layer and the CFTS/Bi$_2$S$_3$ junction, see **Fig. 4.8**, wherein a probe voltage of -0.5 V was given with respect to the ground ITO electrode. The bottom electrode for both the films is ITO whereas the top electrode is the gold-coated SiN. In the present context, the curvature of the *I-V* curve is of major importance. In the figure, the black lines represent forward scans (i.e. from –ve to +ve) whereas the red lines show the backward scans (from +ve to -ve).

Different line types denote the different point of probing on the films. More than fifty different points were scanned on the surface of the film, from what we have selected some representative results in this figure. Clearly, forward and backward scans show similar behavior, resembling a typical diode curve, which is expected from the p-type and n-type nature of the CFTS and Bi_2S_3 layers, respectively. The same thing is also observed in conventional *I-V* characteristics. For the single-layer device **(Fig. 4.8 (a))**, the backward scan differs from the forward one, showing backward curves that indicate charge (minority carriers) accumulation at grain boundaries, limiting the charge transport, and reducing the current. On the other hand **(Fig. 4.8 (b))**, either the majority carriers flow from the n-type Bi_2S_3 across the p-n junction or the attraction of the opposite charge carriers reduces the charge accumulation improves the charge transport. The removal process of excess charge also provides carriers to transport increasing the higher current [156]. That also can be seen in the *I-V* characteristics obtained by C-AFM. It can also be seen that at low bias **(Fig. 4.8(a))**, the backward scan looks like an NDR (negative differential resistance) device [157]. This fact also supports the hypothesis of the accumulation of charge carriers at grain boundaries in single layer CFTS.

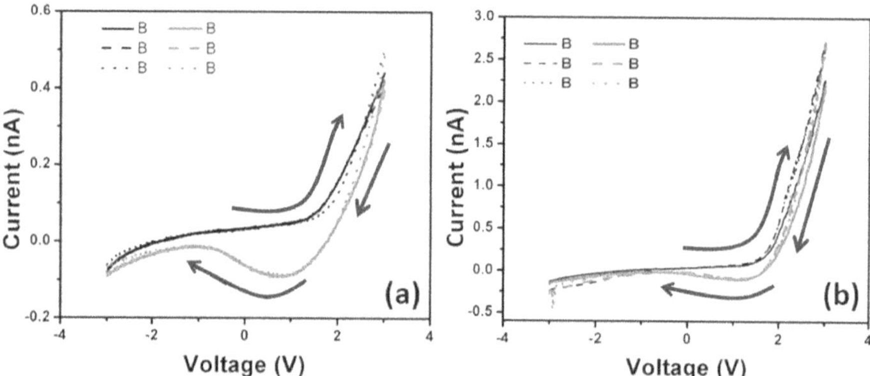

Figure 4.8 *I-V* characteristics of the **(a)** CFTS thin film and **(b)** CFTS/Bi_2S_3 p-n junction taken from conducting an atomic force microscope probe.

(e) Band diagram

A clear picture of this phenomenon has been depicted in a schematic band diagram, **Fig. 4.9**. The picture is drawn referring to the bandgap and conduction and valence band position available in the literature [48]. According to the diagram, the large ITO/CFTS interface barrier is responsible for charge accumulation having non-favorable charge injection type band bending **(Fig. 4.9 (a))**. However, at the junction, the difference between the work functions of the Al electrode and active material has been reduced by inserting the n-type Bi_2S_3 layer **(Fig. 4.9(b))** favoring charge transport and suppressing the electrical bistability [48][158]. In both cases, the release of charge carriers (electrons) is different from one another. In the former case, the charge transfer is difficult from the conduction band (CB of CFTS to Fermi level). Whereas in the latter one, the Bi_2S_3 layer helps in storing the charge or reducing the charge transfer from CB of CFTS to Fermi level.

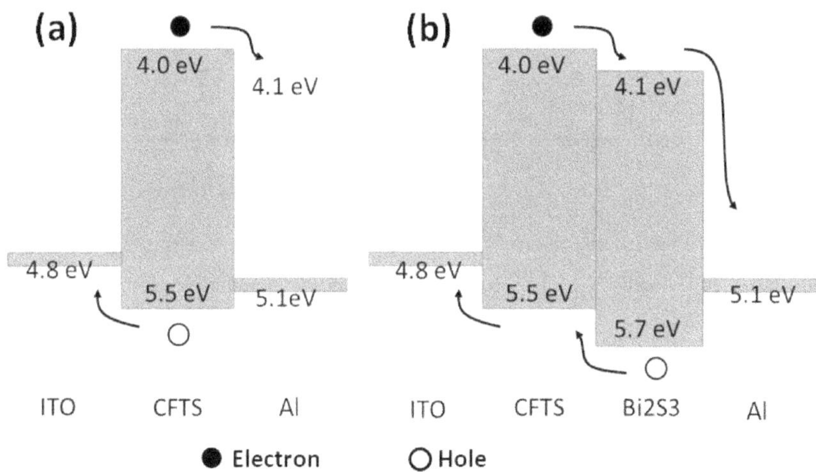

Figure 4.9 A schematic diagrams showing the band alignment of **(a)** ITO/CFTS/Al and **(b)** ITO/CFTS/Bi$_2$S$_3$/Al devices.

(f) Write-read-erase-read sequence

Since the *I-V* characteristics of the CFTS device show both high and low conducting states, we calculated the "write-read-erase-read" sequence of this device as shown in **Fig. 4.10**. Here, pulses of voltage bias of -1 V and 1 V were applied to "write" the high conducting state and then to "erase" to a low conducting state correspondingly. The lower panel of **Fig. 4.10** shows the current value concerning the applied voltage depending upon the preceding "write" or "erase" operation. Additionally, it shows two different values of current for the same probe voltage. The distinction in the current estimations of high and low directing states during testing enables it to be used in various applications like random access memory (RAM) applications.

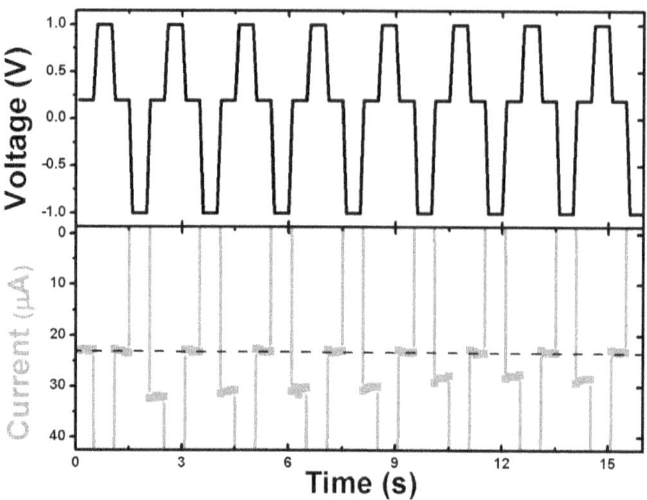

Figure 4.10 Switching performance of a ten layer CFTS device during a continuous "write-read-erase-read" sequence. The voltage pulse is shown in black colour and the current response is shown in red colour.

The electrical bistability in CFTS results shows resistive switching memory phenomena since it has two different current values at one voltage. When a suitable negative bias is applied, firstly the charge carriers that are injected get trapped and assembled to form a conducting path. And when a positive bias is applied then the trapped charge carriers are removed, thereby decreasing the conduction process. Another possibility for the switching behavior may be due to the formation of Cu conductive filament. It was seen that a strong NDR behavior along with resistive switching was observed with excellent repeatability in ITO/CFTS/Al device. But, when the p-n junction device is fabricated then the coexistence of resistive switching and the NDR was not found probably because of the depletion potential formed in the interface which inhibits the formation conducting path.

(f) Electrochemical impedance spectroscopy (EIS)

Electrochemical impedance spectroscopy (EIS) contemplates help in understanding the charge confinement in the CFTS thin film in the high-conducting state and the low-conducting state. Variation of the real (Z') and the imaginary (Z'') parts along with the frequency of the impedance in their high and low-conducting states are shown in **Fig. 4.11**. Here we initially apply a little dc voltage to switch the device to a high-conducting state and immediately measured the impedance followed by applying another suitable dc voltage to make the device operate in the low-conducting state. The high and low-conducting state plots fit a semicircle and the diameter of the semicircle for the high-conducting state decreases due to the switching behavior. Also, the semicircles in the plots **(Fig. 4.11)** confirm that the device can be fitted to a parallel combination of a resistor and a capacitor.

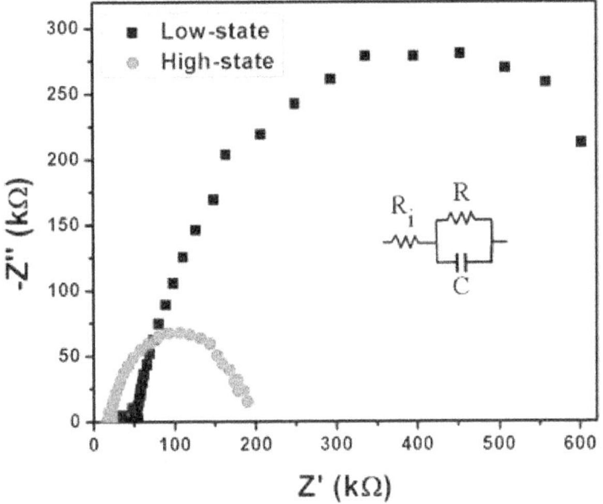

Figure 4.11 EIS of the device after the high-conducting state (red line) and after the low-conducting state (black line) and inset depicts equivalent circuit.

It is evident from **Fig. 4.11**, the semicircle-like nature of the low and high-conducting states of the CFTS thin films, which can be demonstrated to a parallel combination of a resistor and capacitor. In the high-conducting state, a decrease in the resistance is observed due to conductance switching, whereas for the low-conducting state the resistance increases. The plots for this behavior can be visualized in **Fig. 4.12 (a)**, where the resistance values of the CFTS devices are contrived as a function of frequency on a semi-logarithmic scale. Due to switching, the resistance of the devices decreases while increasing the applied frequency. Another key feature of this device is the increase in the conductivity of the device in the high-conducting state due to the induction of the charge in the active layers. Also, we have calculated the capacitance from **Fig. 4.11** (EIS) in the high and low-conducting state as shown in **Fig. 4.12 (b)**. At the high-conducting state, a larger capacitance is seen in the device, revealing that the large amount of charge carriers is trapped during the low-conducting state. Whereas, in the high conducting state the carrier mobility increases which leads to the enhancement in the capacitance during the high conducting state [105]. It is believed that the conductance changes in the device are caused by the polarization due to the applied voltage above a threshold value and making the device to switch to a high-conducting state. Similarly, an appropriate reverse bias is highly essential to restore the device to its low-conducting state.

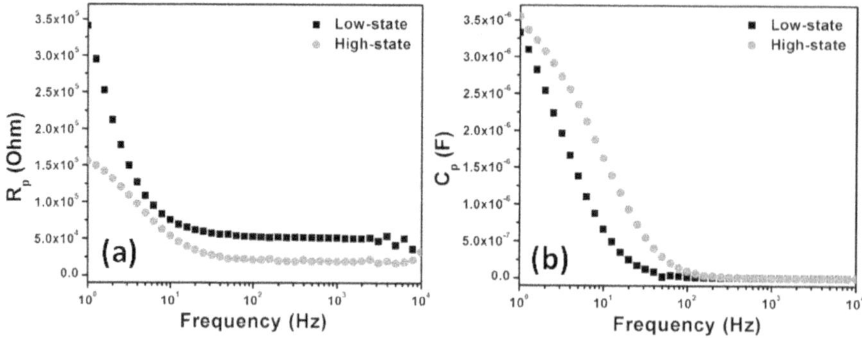

Figure 4.12 Plot for the variation of **(a)** resistance (R_p) and **(b)** capacitance (C_p) of the device based on CFTS thin films in their High and Low-conducting states as a function of frequency.

4.4 Conclusions

In this chapter, we have synthesized CFTS thin films by solution-based SILAR method. The fabricated sandwich device (ITO/CFTS/Al) revealed electrical bistability in the p-type (CFTS) layer with associated memory phenomenon. Experimental studies suggested that the bistability occurred due to charge confinement of the CFTS grained structure on the thin film. The electrochemical impedance investigation of the CFTS gadget uncovers that more charge is stored in the high-conducting state than in the low-conducting state. The high-conducting state demonstrated expanded capacitance and decreased resistance as calculated from the impedance measurements. The device also exhibited random access memory (RAM) behavior demonstrating distinctive current qualities for a similar voltage sequence. The conduction systems of the two conducting states were modeled with the aid of a parallel combination of a resistor and a capacitor.

Chapter 5

Band gap tailoring and change in structural composition within the alloyed semiconductor $Cu_2Fe_{1-x}Ba_xSnS_4$ (CFBTS) via SILAR method and their photoelectrochemical applications

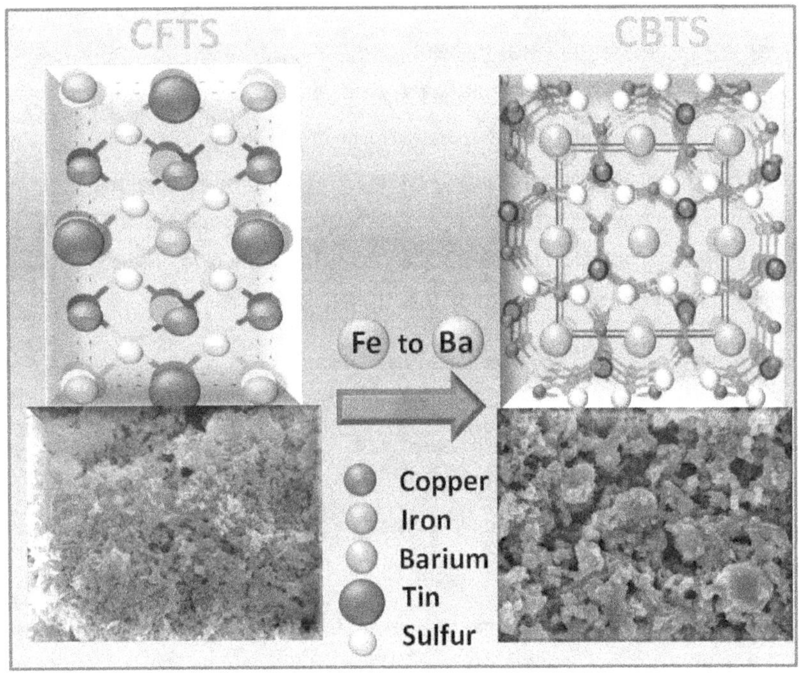

CHAPTER 5

Bandgap tailoring and change in structural composition within the alloyed semiconductor $Cu_2Fe_{1-x}Ba_xSnS_4$ (CFBTS) via SILAR method and their photoelectrochemical applications

5.1 Introduction

In the previous chapter, we were able to synthesis pure phase CFTS thin film with the help of the SILAR method. Hence we will proceed with the SILAR method in this chapter also due to its excellent film formation properties as mentioned in the previous chapters. So here we use this technique to partially substitute Ba to replace Fe to form $Cu_2Fe_{1-x}Ba_xSnS_4$ (CFBTS). So a gradient band gap absorber film with continuously tunable bandgap is demonstrated in this chapter with the help of a solution-processed SILAR method. Also, some of the previously mentioned synthesis techniques involve the use of toxic organic solvents and stabilizers [78].

SILAR method is a promising large-scale film fabricating method that is used in this work for obtaining excellent film quality and is much of a low-cost technique [105]. For most of the compound semiconductors, the improvement in efficiency is controlled usually by precisely adjusting the annealing temperature usually termed as sulfurization [133, 159]. And also, by adjusting the ratio between the cations partially or completely we can tune the bandgap of the semiconductor. However, by adjusting the anion ratio of selenium to sulfur, we can tune the bandgap and thereby increasing the efficiency [160, 161].

A narrower bandgap can be obtained by replacing the heavier atoms from the same group in the periodic table for these semiconductors. The bandgap of CFTS lies between 1.26 to 1.50 eV and it needs to be reduced to around 1.10 to 1.25 eV to obtain optimum absorption and higher efficiency [116]. As the bandgap of CZTS (1.5 eV), can be reduced by partially substituting a larger atomic size cation such as cadmium and thus obtaining a reduced bandgap of 1.41 eV [19].

On the other hand, the bandgap of kesterite CZTS can be increased as well by the partial substitution of Ba in $Cu_2Zn_{1-x}Ba_xSnS_4$ (CZBTS). Whereas, Cu_2BaSns_4 (CBTS) is an earth-abundant wide bandgap semiconductor which can be used as top cell for photovoltaic applications due to its better optoelectronic and defect properties. It also has high optical

100

absorption coefficient in the visible light range because of high joint density of states. Thus, CBTS exhibit lower density of non radiative recombination centres [38].Therefore, the bandgap tuning can be achieved by the partial substitution of various cations accordingly. Recently, Xiao et al. recommended distant-atom mutation with a group (II) like Barium which is located far off in the periodic table [34].

Hence, the changeable electronic properties and large ionic size along with charge mismatch among the constituents of Fe (1.26 Å) and Ba (2.68 Å) are free from forming any kind of cation-cation disorder and was theoretically proposed as light-absorbing material [162]. Furthermore, partial substitution of Ba in stannite CFTS is still unexplored. So, we try to substitute Fe partially with Ba in CFTS to form $Cu_2Fe_{1-x}Ba_xSnS_4$ (CFBTS). Hence, by using a larger Ba atom we can replace Fe from its lattice and can increase the bandgap followed by a reduction in the antisite defect [163].

CBTS has been used as a light absorber for both photovoltaic as well as photoelectrochemical (PEC) applications. CBTS has a large chemical potential window because of large ionic size due to which it is more thermodynamically stable than that of CFTS [37]. For PEC applications, CBTS thin films were fabricated with few over layers like CdS/ZnO/TiO$_2$ on CBTS thin films as reported by Ge et al. with 7.5 mA cm^{-2} at 0 V vs RHE [38]. Whereas, Zhou et al. recently reported the most efficient and stable photocathode with CBTS/CdS/TiO$_2$/Pt architecture generating a photocurrent of 12.08 mA cm^{-2} [39]. Despite these remarkable growths in the history of CBTS, most of the CBTS thin films have been fabricated with over layers and vacuum-based techniques. Thus, it is required to develop a low-cost solution-processed method for CBTS thin film fabrication for PEC application in the production of solar hydrogen.

In this regard, here we report an original study on the structural changes of highly crystalline CFTS thin films by a low cost and simple successive ionic layer and adsorption reaction (SILAR) technique. Followed by partial substitution of Fe by Ba to fabricate CFBTS thin films where x varies from 0, 0.25, 0.50, 0.75 to 1 and labelled as CFTS, B25, B50, B75, and CBTS respectively. Finally, CBTS thin film coated on Mo substrate was fabricated, which was then sulfurized to make devoid of impurity peaks. This study was investigated with a basic structure consisting of Glass/FTO/CFBTS without any over layers and the photoresponse of these CFBTS thin films was investigated by fabricating PEC.

5.2 Experiment

5.2.1 Chemicals and materials required

Copper chloride dihydrate (99%, Alfa Aesar), iron nitrate nonahydrate (99%, Thomas Baker), barium chloride (99%, Thomas Baker), tin chloride (99.9%, Sigma Aldrich), sodium sulfide (99%, Merck), 2-methoxy ethanol (99%, S.D. Fine chemicals), deionized water (DI water), Fluorine tin oxide (FTO, 8-12 Ω/sq, BAT-SOL, Australia), aluminium pellet (3-5 mm, 99.99%, Sigma-Aldrich, Germany). All these materials were used as bought without any further purification

5.2.2 Synthesis of CFBTS thin films by SILAR method

First, we prepared a CFTS thin film in the molar ratio 0.2:0.1:0.1:0.4 by the SILAR method. The cationic precursor solutions were prepared in separate beakers and finally, all of the cationic precursor solutions consisting of $CuSO_4.5H_2O$, $Fe(NO_3)_3.9H_2O$, and $SnCl_2.2H_2O$ in 3 mL ethanol each, in a stoichiometric molar ratio of 0.2:0.1:0.1 were mixed and stirred for half an hour. Initially, $Fe(NO_3)_3$ solution was added to the $SnCl_2$ solution which resulted in a golden yellow colored solution, followed by the addition of $CuSO_4$ solution resulting in a greenish-yellow solution. A separate beaker with 0.4 M Na_2S in 9 mL ethanol was used as the anionic precursor solution. Then we synthesized the four different types of CFBTS (B25, B50, B75, and CBTS) by substituting Ba with Fe.

The SILAR solutions were prepared by dissolving $Cu_2SO_4.5H_2O$, $Fe(NO_3)_3.9H_2O$, $BaCl_2.2H_2O$, $SnCl_2.2H_2O$ in 3 mL, and Na_2S in 9 mL ethanol each. Then for fabricating CBTS thin films, precursor solutions were prepared by $Cu_2SO_4.5H_2O:BaCl_2.2H_2O:SnCl_2:Na_2S$ in the same 0.2:0.1:0.1:0.4 ratios respectively as done for CFTS. To minimize the oxidation the excess amount of sulfur is used to minimize the oxidation and to prevent the sulfur deficiency in the as-prepared thin films. The prepared solutions were coated on soda-lime glass (SLG) and fluorine tin oxide (FTO) glass substrates using the SILAR technique. The SLG and FTO substrates were cleaned twice by successive ultrasonic treatments in acetone, ethanol, and distilled water for 10 min each.

The SILAR technique was performed in four beaker systems which include various experimental parameters, such as plunging time, number of cycles, and precursor concentration. The pre-cleaned substrates were immersed in a cationic precursor solution in the first beaker for 30 s to adsorb ions on the glass substrate. The substrates were then rinsed with ethanol in the second beaker for 15 s to eliminate any loosely bound ions. Subsequently, the substrates were then immersed for 30 s in an anionic precursor solution in the third beaker where the S^{2-} reacts with the cations to form the CBTS layer on a glass substrate. At last, the substrates were altogether thoroughly rinsed with ethanol in the fourth beaker for 15 s to remove excess ions and powdery precipitate from the surface of the film. A total of ten numbers of cycles were coated on both the substrates using the SILAR method and heated at 150 °C after the ten cycles. Then, CBTS thin film was coated on Mo substrate and compared it with the sample coated on FTO by the SILAR method. The CBTS sample coated on Mo substrate was subjected to sulfurization at 500 °C for half an hour to study the PEC measurement changes after sulfurization.

5.2.3 CFBTS thin film characterization

The as-prepared thin films were characterized using the Ultima IV model X-ray diffractometer equipped with a Cu Kα source (λ=1.54183 Å) to study the crystal phase and structural properties. Additional characterization was done by recording the Raman spectra using Horiba Raman spectroscopy (excitation wavelength 532 nm). The morphological and compositional analysis of our CFTS thin films were characterized by energy-dispersive X-ray spectroscopy (EDS) equipped in a field-emission scanning electron microscope (FE-SEM) (Hitachi S4800 FE-SEM). UV-visible absorption spectroscopy using Lambda 950- Perkin Elmer instrument was done to study the optical properties such as a change in the band gaps of the fabricated films using the SILAR method. The photoresponsive behaviour of the as-prepared thin films was measured under dark and light illumination using Keithley 2450 source meter simulated at AM 1.5G solar illumination.

5.2.4 Three electrode system for device testing

The photoelectrochemical (PEC) measurements were carried out by potentiostat Biologic work station model no. SP150 and the standard 3-electrode cell configuration using an Ag/AgCl with 1 M KCl as reference electrode and with platinum mesh as the counter electrode. A scan rate of 5 mV s^{-1} was applied to acquire the data. And an irradiation intensity of 100 mW cm^{-2} was illuminated onto the photocathode. A neutral electrolyte solution having pH ~7 of 0.1 M sodium sulfate was used for all the PEC measurements. We measured as-prepared CFTS, B25, B50, B75, CBTS, and sulfurized CBTS samples were tested by the PEC method. During the measurement, the as-prepared CFTS, B25, B50, B75, CBTS, and sulfurized CBTS samples served as the working electrodes. The potential versus reversible hydrogen electrode (E_{RHE}) was converted from potentials versus Ag/AgCl electrode ($E_{Ag/AgCl}$) by using the Nernst equation, $E_{RHE}= E_{Ag/AgCl}+0.059\times pH0.201$.

5.3 Results and Discussion

5.3.1 As-prepared CFBTS nanoparticles

5.3.1.1 Characterization of as-prepared CFBTS nanoparticles

(a) X-ray diffraction (XRD)

The crystal structure, structural properties, and the phase identities of the as-fabricated $Cu_2Fe_{1-x}Ba_xSnS_4$ (CFBTS) thin films were determined by the XRD pattern analysis. The XRD patterns of the CFTS, B25, B50, B75, and CBTS are shown in **Fig. 5.1**. The strong diffraction peaks at $2\theta=28.46°$, $44.35°$ and $56.38°$ for CFTS and at $2\theta=27.72°$ and $28.01°$ for CBTS, corresponds to diffraction planes 112, 220, 312 of polycrystalline stannite CFTS (PCPDF: 17-0320) and diffraction planes 104 and 110 of trigonal CBTS phase (PCPDF: 97-005-2685) respectively [162, 163]. As illustrated in **Fig. 5.1 (b)**, the prominent diffraction peak 112 of as-fabricated CFBTS (0<x<1) thin films was found to be slightly shifted to the lower 2θ angle with increasing Ba content.

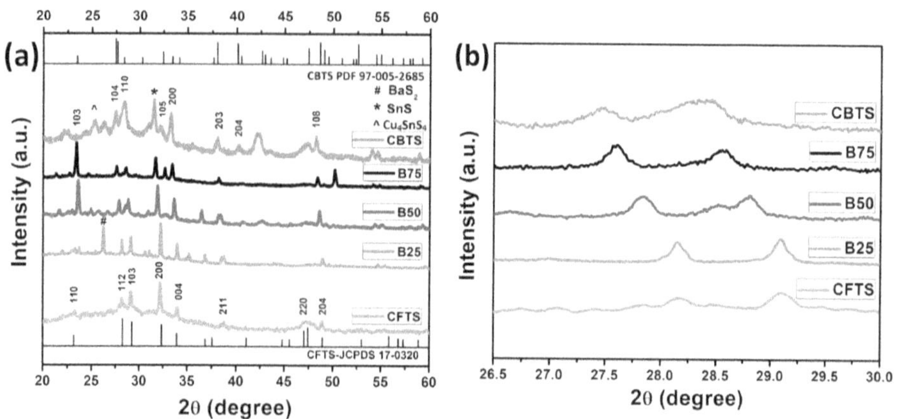

Figure 5.1 (a) XRD patterns of CFTS, B25, B50, B75 and CBTS thin films and **(b)** Magnified peaks of 104 and 110 planes.

This shift of the dominant peak is attributed to the expansion of CFTS unit cell volume with increasing Ba content as bigger Ba^{2+} (0.268 nm) replaces Fe^{2+} (0.126 nm), which can weaken the anti-bonding component between Sn and S in CFBTS. The dominant shift observed with increasing Ba content in CFBTS quinary alloy nanostructures exhibits a phase transition from stannite to trigonal. The shifting of 112 and 103 peaks of CFBTS are shown in **Fig. 5.1 (b).** Also, it was observed that the grain size and Ba content in CFBTS alloy nanostructures vary as a direct relation. This is evident from the scanning electron microscope (SEM) images of the thin films as shown in **Fig. 5.1**. However, along with the main peaks of CFBTS, few secondary peaks such as BaS_2 and Cu_4SnS_4 peaks are also observed in the CBTS thin films. Also, an SnS peak at 31.5 °C is labelled with a star mark, which has also been found by other groups [164].

(b) Raman spectroscopy

Since the structural change from the stannite and trigonal phase results in the formation of a few secondary phases which may not be identified with the help of XRD alone due to the very narrow single-phase region. Generally, the Raman spectroscopy is effectively used to identify these coexisting secondary phases and to detect the vibration modes of CFBTS thin films. The CFBTS films exhibit the strongest peaks from 317 to 336 cm^{-1} (when x=0 to 1) for all the five samples which can be attributed to the symmetric vibrations of the motion of sulfur atoms as shown in **Fig. 5.2 (a)**.

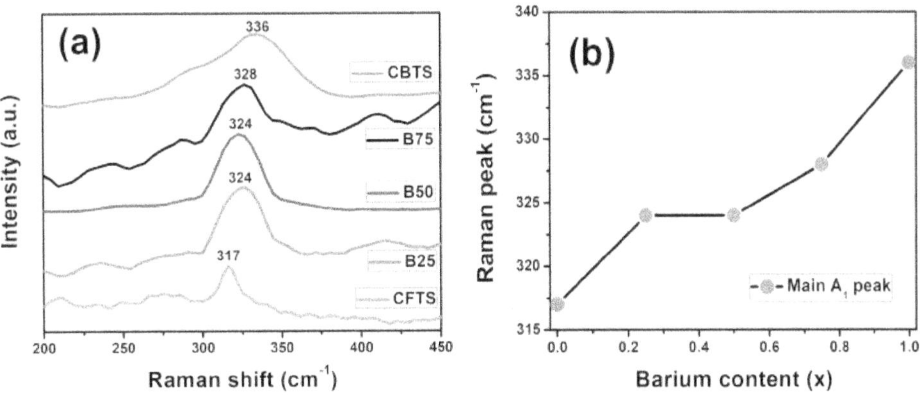

Figure 5.2 (a) Raman spectra of CFTS, B25, B50, B75 and CBTS thin films and **(b)** variation of Raman peak vs. barium content.

The SnS$_2$ and Cu$_2$SnS$_3$ impurity peaks were also detected from the Raman spectra. The strong vibration intensity of the Raman peaks indicates that the as-grown films are highly crystalline in nature. The peak position of A$_1$ mode with varying Ba content in CFBTS is shown in **Fig. 5.2(b)**. Also, the position of the A$_1$ mode is towards the high-frequency direction with the increase in the Ba content. This increase is caused due to the increase in the bond stretching force constants.

(c) Field emission scanning electron microscopy (FE-SEM)

Here for the first time, CFBTS thin films were prepared by a simple SILAR method on the fluorine-doped tin oxide (FTO) coated soda-lime glass (SLG) substrates without any high-temperature annealing or sulfurization. The FESEM surace images of CFTS, B50, and CBTS thin films are shown in **Fig. 5.3**. The films are unevenly distributed but agree with the stoichiometric ratio of the elements. **Fig. 5.3 (a)** shows that the as-prepared CFTS thin film exhibited smaller nanoparticles without any grains when compared with other samples. Whereas, the substitution of Ba in B50 and CBTS resulted in the increased grains thereby forming large CFBTS grains extending throughout the entire layer thickness as shown in **Fig. 5.3 (b and c)**.

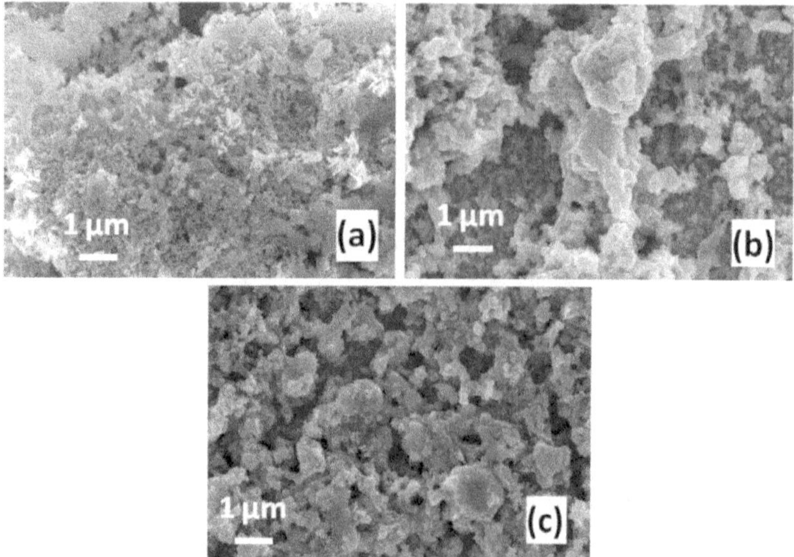

Figure 5.3 FE-SEM images of the Cu$_2$Fe$_{1-x}$Ba$_x$SnS$_4$ **(a)** CFTS, **(b)** x=B50 and **(c)** CBTS thin films grown on FTO coated glass substrates.

The substitution of the Ba resulted in the nucleation and subsequent crystal growth processes compact grain formation. It can be seen that the CFTS thin film is free from pin-holes and is compact. Whereas, the other two thin films (B50 and CBTS) have pin-holes and are not compact when compared with the CFTS sample. This can be further proved with the help of SEM-EDX analysis, as it will identify the chemical composition and percentage of elements present. Also shows the formation of large CFBTS grains and thus these micron-sized grains are useful for fabricating photovoltaic devices. As these larger grains often exhibit longer excited-state lifetime, lower trap density, and relatively less open-circuit voltage (V_{oc}) loss with higher PCE.

(d) Energy dispersive X-ray spectroscopy (EDS)

To understand the elemental distribution in the CFBTS thin films, energy dispersive X-ray (EDX) analysis was performed. The SEM-EDX elemental mapping of Cu, Fe, Ba, Sn, and S of CFTS, B50, and CBTS thin films are shown in **Fig. 5.4.** In CFBTS, where CFTS demonstrates a Fe deficient and slightly Cu rich composition, whereas CBTS thin films match well with the stoichiometric ratio of the elements as shown in elemental compositional EDX analyses in **Table 5.1**.

Table 5.1 Elemental compositions of CFTS, B50 and CBTS thin films for different Ba contents.

Samples	Chemical Composition (At.%)					Composition Ratio	
	Cu	Fe	Ba	Sn	S	Cu/(Fe+Ba+Sn)	S/metal
CFTS	27.59	8.80	0	14.14	49.47	1.20	0.97
B50	33.75	3.77	3.81	12.09	46.58	1.71	0.87
CBTS	26.73	0	12.08	11.18	50.01	1.14	1.00

Additionally, Cu rich composition can be attributed to grains with better transport. Also, solar cells fabricated under this condition usually exhibit strong interfacial recombination which can effectively decrease the minority charge carrier lifetime and thus hinders the device efficiencies, which limits the device performance. However, etching with an aqueous solution of KCN can effectively increase the device efficiency by selectively removing certain secondary phases [165]. The possible secondary phases are Cu_XSe and the excess chalcogen on the surface. While the B50 exhibited an almost equal amount of Fe and Ba content but rich in Cu content. While **Fig. 5.4** shows the SEM images of the B50 thin film and SEM-EDX elemental mapping images depicting the elements uniformly distributed Cu, Fe, Ba, Sn and S in red, yellow, green, blue, and purple respectively are shown in **Fig. 5.4 (b-f)**. The above characterizations and parameters of the as-grown CFBTS thin films are excellent for the fabrication of high-performance semiconductor devices.

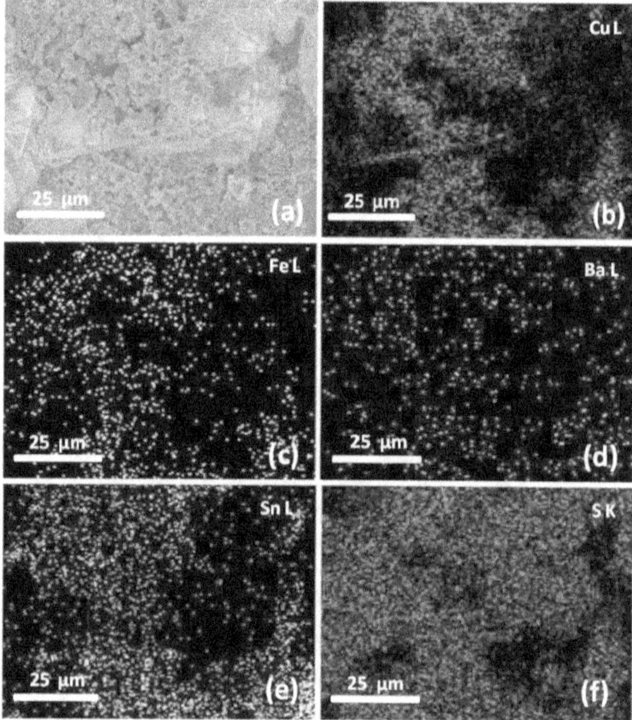

Figure 5.4 (a) The SEM image of B50 thin film and **(b-f)** SEM-EDX of elemental mapping images of Cu, Fe, Ba, Sn, and S in red, yellow, green, blue, and purple respectively are showed.

(e) UV-Vis spectroscopy

To study the influence of substitution of Ba on the bandgap of CFBTS, the films were measured by ultraviolet-visible (UV-Vis) spectroscopy. This optical characterization proves that the films are potential candidates for the photoelectrochemical (PEC) applications. CFTS and CBTS exhibit light absorption at a wide range in the visible region. The bandgap of CFBTS thin films are calculated using the tauc plot by extrapolating the linear region as shown in **Fig. 5.5**.

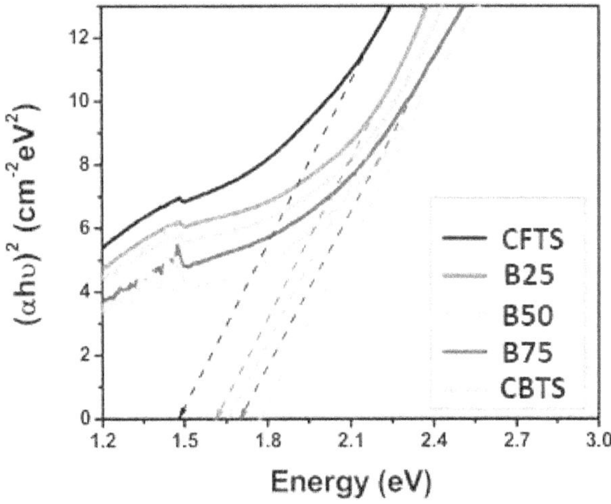

Figure 5.5 The bandgap of CFTS, B25, B50, B75 and CBTS thin films.

As the Ba content increases there is a linear increase in the bandgap is observed. The dependence of band gaps with the ratio of Ba/(Fe+Ba) for $Cu_2Fe_{1-x}Ba_xSnS_4$ thin films is shown in **Fig. 5.6**. When the Ba content increases from 0 to 1.0, the band gaps of CFBTS thin films increases from ~1.5 eV (CFTS) to ~1.8 eV (CBTS). This is considered as the optimum bandgap required for fabricating a solar cell with higher power conversion efficiency. However, the first reported prototype solar cell with the pure sulfur compound was 2.02 eV and yet yields lower efficiency [35, 166]. Moreover, based on the XRD of the CFBTS thin films, the structural change from stannite CFTS to trigonal CBTS phase can also be explained by the change in the band gaps for CFBTS thin films while increasing the Ba content.

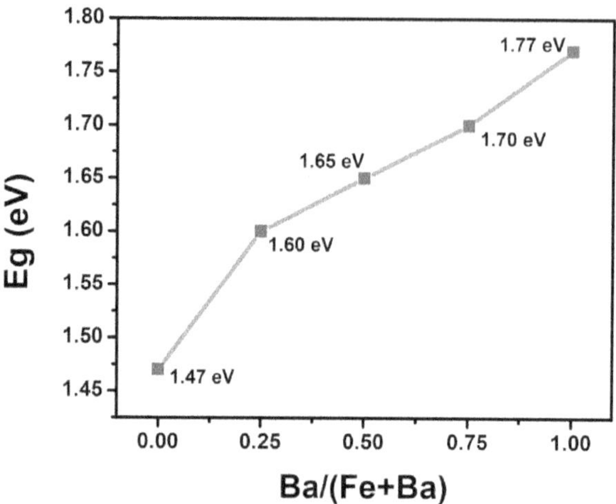

Figure 5.6 The dependence of band gaps with the ratio of Ba/(Fe+Ba) for $Cu_2Fe_{1-x}Ba_xSnS_4$ thin films.

(f) *I-V* **characteristics of CFBTS thin films**

To find out the potential use of these CFBTS thin films in photovoltaic applications as an absorber layer we have studied the photoresponse of CFTS and CBTS thin films as shown in **Fig. 5.7** and B25, B50, and B75 CFBTS thin films are depicted in **Fig. 5.8**. The photoresponse of these devices is characterized by the aid of FTO we use aluminium (Al) as back contact and as top contact we use (Glass/FTO/CFBTS/Al). Current-voltage (*I-V*) curves of these thin films were measured both in dark and under light illumination using a solar simulator with AM 1.5 G and 100 mW/cm^2 intensity. It is observed from the *I-V* characteristics that the value of current has tremendously increased from 0.4 to 27 µA for CFTS to CBTS respectively with an obvious photoresponse property. The higher current value indicates that the effective charge transfer characteristics as and when the Fe is substituted by Ba. Similarly, with increasing Ba content the current value is being increased linearly in the sequence CFTS<B25<B50<B75<CBTS thin films. When the light is incident, the electrons from the valence band to the conduction band get excited, thereby increasing the hole concentration in the valence band which further enhances the current in the thin films. This phenomenon validates the generation of photocurrent in thin films and finds potential in photovoltaic applications.

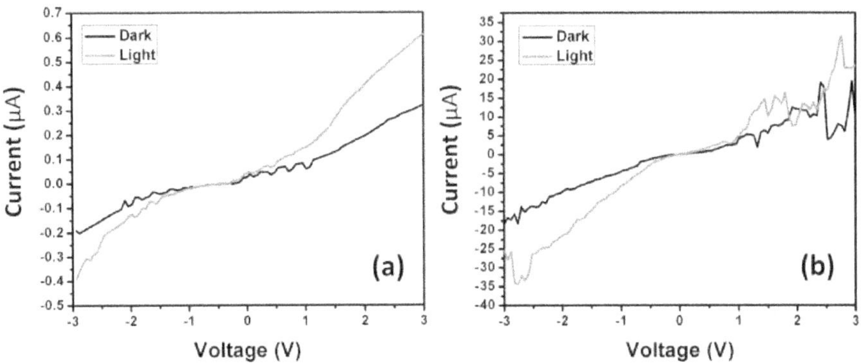

Figure 5.7 *I-V* characteristics of **(a)** CFTS and **(b)** CBTS thin films under AM 1.5G simulated solar illumination.

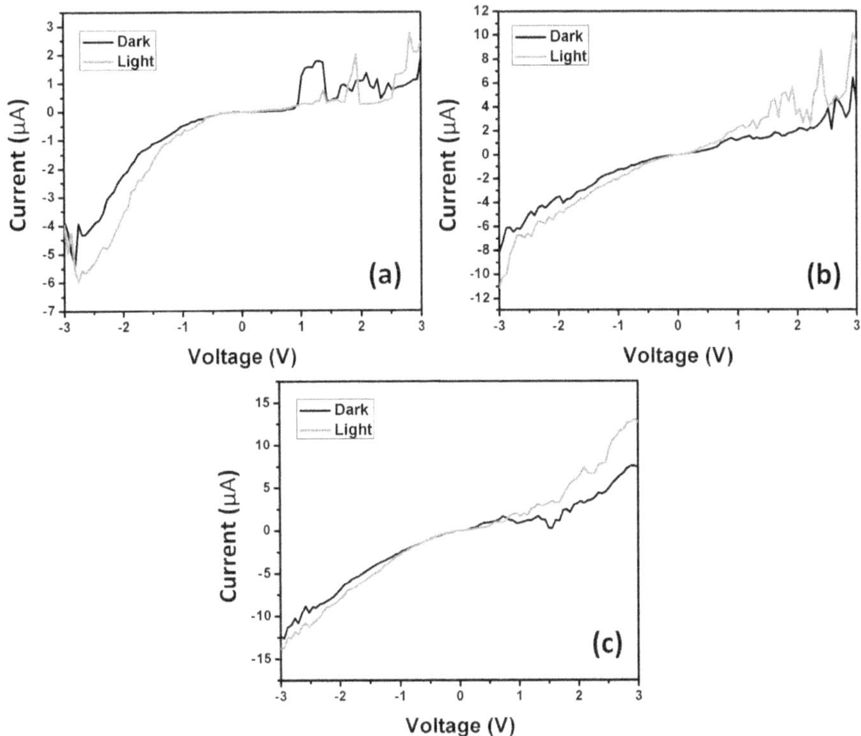

Figure 5.8 *I-V* characteristics of **(a)** B25, **(b)** B50, and **(c)** B75 thin films under AM 1.5G simulated solar illumination.

(g) Electrochemical impedance spectroscopy (EIS)

The bare CBTS thin films coated on FTO substrate which exhibited high photocurrent was then investigated with the help of electrochemical impedance spectroscopy (EIS) to study improved charge collection and transportation is shown in **Fig. 5.9**. EIS shows the Nyquist plot measured in dark and simulated solar light illumination for the bare CBTS sample. It was observed that the radius of the semicircle region in the dark condition has increased when compared with light illumination. This can be attributed to the reduction in the interfacial charge transfer resistance between the CBTS and FTO electrodes. The semi-circular portion in the high-frequency region corresponds to the charge transfer process in the electrolyte-electrode interface. Thereby, confirming that the bare CBTS film after light illumination lowers the resistance of the interfacial charge transfer between the working electrode and FTO substrate. The linear component present at the lower frequency region reveals the diffusion process of ions in the electrolyte occurring in the proximity of the semiconductor-electrolyte interface. The straight line at low frequency is due to Warburg impedance (Z_w) which is used to model the ionic diffusion of the electro active species. The values of series resistance associated with the electrolyte (R_s), the charge transfer resistance (R_{ct}), the capacitance of electrical double layer (C_d) and Z_w can be calculated by fitting the data into equivalent circuit as shown in the inset of **Fig. 5.9**.

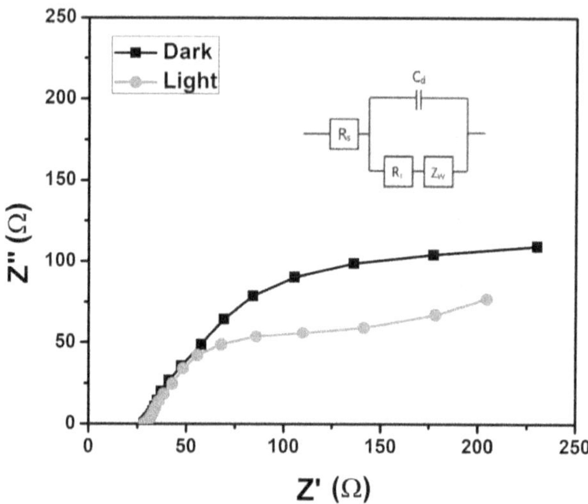

Figure 5.9 Nyquist plot of as-prepared CBTS thin films measured in dark and under light illumination and the inset shows the equivalent circuit.

5.4 Sulfurized CBTS thin films

(a) X-ray diffraction (XRD)

To analyse the structure and formation mechanism of CBTS thin films, the XRD patterns of the CBTS films with and without sulfurization process are collected. The CBTS thin film was sulfurized at 500 °C for one hour in the presence of sulfur powder under flowing argon gas. After sulfurization, XRD pattern of the sulfurized CBTS thin film exhibited a pure stannite phase without any impurities and it was plotted along with the XRD pattern of as-prepared CBTS thin film for comparison which resulted in few secondary peaks as shown in **Fig. 5.10**. The XRD patterns show high crystallinity of the films with all the peaks assigned with the trigonal phase. The sulfurization process led the samples to show strong diffraction peaks are well indexed to the JCPDS card number 97-005-2685 of the trigonal CBTS and except for the diffraction peak of Mo marked at 40.5 °C marked as "*".

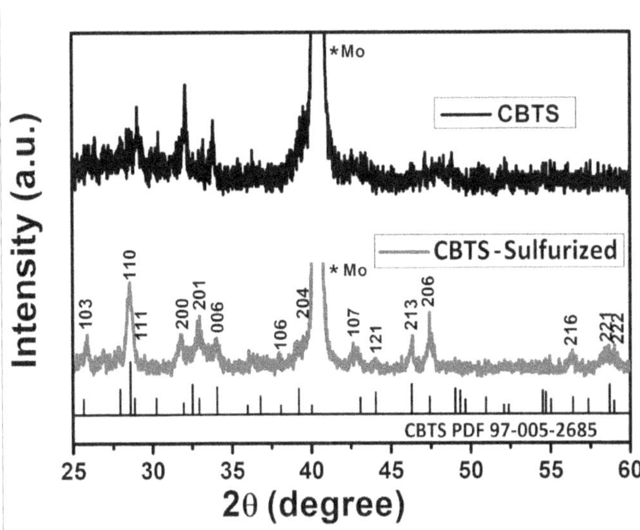

Figure 5.10 XRD patterns of as-prepared CBTS (black) and sulfurized CBTS (blue) thin films.

(b) Field emission scanning electron microscopy (FE-SEM)

Fig. 5.11 shows the top-view SEM images of sulfurized CBTS films deposited on a Mo substrate. It can be seen that the sulfurized CBTS films deposited by the solution-processed SILAR method are free from pinholes and compact without any secondary phases with the formation of larger grains compared to as-prepared CBTS thin films. Fig. 5.11 (b) shows the magnified SEM image of sulfurized CBTS films with contiguous and uniform without any pinholes which are most important parameters for a better performance CBTS PEC device. It also indicated that the formation of high purity and uniform morphology CBTS phase after sulfurization process. This sulfurization process helps in increasing the grain size growth thereby reducing the recombination centres located at the grain boundaries. The SEM result shows that the sulfurization can successfully prevent the secondary phase formation, which results in CBTS films with improved crystallinity and uniformly distributed grains.

Figure 5.11 FE-SEM images of the **(a)** sulfurized CBTS and **(b)** magnified sulfurized CBTS thin films.

(c) Electrochemical impedance spectroscopy (EIS)

The sulfurized CBTS thin films coated on Mo substrate which exhibited high photocurrent was then investigated with the help of electrochemical impedance spectroscopy (EIS) as shown in **Fig. 5.12**. EIS provides the impedance response of the sample against wide frequency range. There is a small semicircle in the high frequency region and a straight line in the low frequency region. The arc at the high frequency impedance region should stem from the capacitance of the space charge region and its accompanying resistance element due to the contact of the electrolyte and semiconducting material. The linear component at the low frequency region shows the diffusion of ions in the electrolyte occurring in the vicinity of semiconductor electrolyte interface. Warburg element can be used to model the ionic diffusion in the low frequency region. The impedance arc has become smaller, which is due to sulfurization process that resulted in either reduction of defects and thereby reducing the charge transfer resistance and it is shown by the electrical circuit fitted by EIS data.

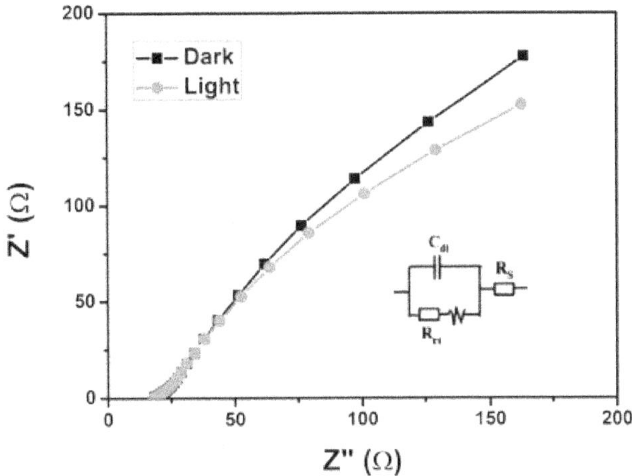

Figure 5.12 Nyquist plot of sulfurized CBTS thin films measured in dark and under light illumination and the inset shows the equivalent circuit.

5.5 Photoelectrochemical (PEC) studies

(a) Linear sweep voltammetry (LSV)

To further study the photoconductivity of the SILAR grown CFBTS thin films, photoelectrochemical (PEC) studies were directed. A five different photocathode with a configuration of Glass/FTO/CFBTS were fabricated using CFTS, B25, B50, B75, and CBTS absorber layers as working electrodes. The platinum mesh was used as a counter electrode, Ag/AgCl was used as a reference electrode and 0.1 M Na_2SO_4 was used as electrolyte respectively. Basically, in PEC the electrolyte acts as a charge transfer between the working electrode and counter electrode and it is based on the junction formed between the semiconductor material and electrolyte. In **Fig. 5.13 (b)** the linear sweep voltammetry (LSV) measurements of the Glass/FTO/CBTS/electrolyte PEC cell measured in both dark and light illumination conditions are seen. At first, bare CFTS on the FTO substrate was coated and tested in which it generated a low amount of cathodic photocurrents under the illumination of 100 mW cm^{-2} xenon lamp as shown in **Fig. 5.13 (a)**. As and when the Ba content increased the photocurrent was increasing in B25<B50<B75 trend, but the difference in the dark and light was less as shown in **Fig. 5.14 (a-c)**. Whereas, bare CBTS exhibited enhanced photocurrents under illumination which indicates that the bare CBTS can be used as a photocathode and has a p-type conductivity as shown in **Fig. 5.13 (b)**. The PEC cells based

on CBTS films deliver a photocurrent density difference of 0.68 mA cm^{-2}. The scan rate was set at 5 mV s^{-1} during linear sweep voltammetry in this work. The photocurrent density difference of the sulfurized films was much better than the as-prepared samples (0.75 mA cm^{-2}) as shown in **Fig. 5.13 (c)**. Also, the overall current density has been increased after sulfurization. The increase in the photocurrent after introducing Ba in CFBTS thin films is attributed to the process in which the electrons get transferred from valence band to conduction band when excited by incident light. These electrons are then transferred to the CBTS electrode and electrolyte interface. Depending on the recombination effect of electrons and holes, and its path length decides whether the photo excited electrons can take part in a water reduction reaction for hydrogen evolution. As all the device structure under the light illumination showed enhancement in the current density. Thus, these copper-based chalcogenide thin-film devices can be used in PEC applications.

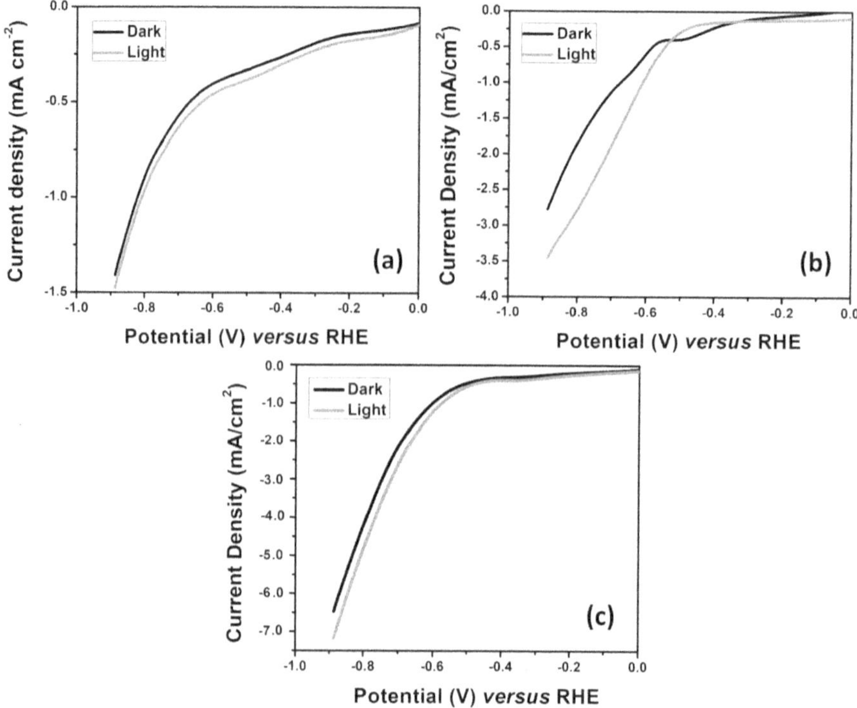

Figure 5.13 Photoelectrochemical *J-V* characterizations under dark and light illumination of (a) CFTS (b) CBTS and (c) Sulfurized CBTS thin films.

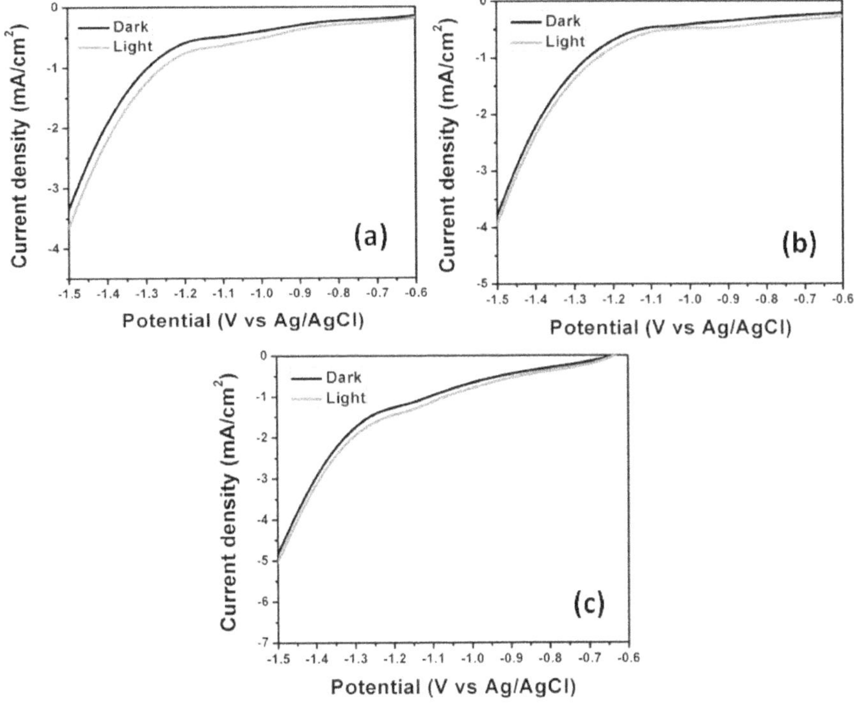

Figure 5.14 Linear sweep voltammetry (LSV) curves of (a) B25, (b) B50 and (c) B75 PEC cells.

(b) Chronoamperometry

Chronoamperometry studies are conducted for the as-prepared and sulfurized CBTS thin films as shown in **Fig. 5.15**. Chronoamperometry is used to study the diffusion process, adsorption of charge particles and kinetics of a chemical reaction. In this technique, a potential step is applied to the electrode and the resulting current versus time is observed. This can be analysed by manually chopping the light. When the illumination was turned on, the rapid increase in the cathodic photocurrent was observed for both as-prepared and sulfurized CBTS thin films. Similarly, when the light illumination was turned off, it took longer time for the photocurrent decay to occur which shows that the persistent nature of the photoconductivity. However the sulfurized CBTS thin film resulted in more current generation compared with as-prepared CBTS sample. These types of prolonged decays of photocurrents are observed in the as-prepared and sulfurized CBTS thin films as well as various other photoactive materials. Persistent photoconductivity is a light-induced enhancement in the conductivity that persists for a long period of time after the light switch off is arisen because of electrostatic potential fluctuation in the band structure and defect related to the secondary phase in the structure. Thus the SILAR deposited CBTS thin films need to have various over layers to make it defect free and for high efficiency PEC studies.

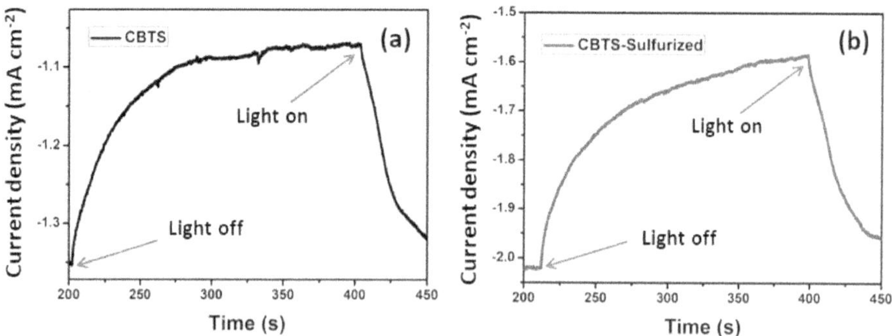

Figure 5.15 Photoelectrochemical chopped illumination of **(a)** CBTS and **(c)** sulfurized CBTS thin films.

5.6 Conclusions

In conclusion, we have successfully synthesized polycrystalline $Cu_2Fe_{1-x}Ba_xSnS_4$ (CFBTS) alloy thin films using the SILAR method without any post-annealing process. The X-ray diffraction patterns exhibited the structural transition from stannite to trigonal phase with the increasing barium content. The band gaps of the as-prepared CFBTS thin films were found to be increased from ~1.5 to ~1.8 eV, almost in a linear increasing trend with the increasing barium content. Finally, a high performance photoelectrochemical (PEC) cells Glass/FTO/CFBTS photocathode with improved stability was fabricated by a single step SILAR method. The sulfurized CBTS thin film exhibited phase pure trigonal structure with an improved photocurrent density difference of 0.75 mA cm^{-2}. Without the addition of any bilayers on the CFBTS, a significant improvement of photocurrent and onset potential was observed in the CFBTS thin films. Also, we measured the photoresponse of these CFBTS thin-film alloys which confirms that it can be utilized to fabricate solar cells with high power conversion efficiency.

Chapter 6

Overview of the Present Work and Future Prospects

CHAPTER 6

Overview of the Present Work and Future Prospects

6.1 Overview

In the present work, the synthesis of CFTS nanoparticles with the help of different methods is investigated in detail such as solvothermal route, sol-gel, and SILAR method. Also, multifunctional applications of CFTS thin films are elaborated in this thesis. CFTS were initially fabricated with the aid of PEG with varying chain lengths as solvent followed by sulfurization. This synthesis is followed by the focus on the photo-enhanced supercapacitive behavior of photoactive (CFTS) nanoparticles. Then fabrication of different types of CFTS thin films with the help of the sol-gel method and the spin coating method was adopted. The effects of various stabilizers in CFTS thin film fabrication were studied, in which different stabilizers like lactic acid and monoethanolamine were administered to the CFTS sol solutions. This was done to check whether this helps in reducing the secondary phase and then it was followed by the sulfurization process as the formation of the required phase wasn't completed. Followed by this work, the CFTS thin films were fabricated with the help of a solution-based SILAR method which resulted in a pure stannite phase. These CFTS thin films were used for the study of the co-existence of negative differential resistance and resistive switching in-memory applications. Then the effect of the p-n junction was explored by incorporating an n-type Bi_2S_3 thin film fabricated besides the p-type CFTS thin film. Finally, the substitution of Zn with Fe was not sufficient to increase its optical properties and electrical properties, therefore the partial substitution of Fe with Ba in the CFTS compound semiconductor was conducted by the SILAR method to find out how the optical and other properties vary. And the physical, chemical, and electrical properties of polycrystalline $Cu_2Fe_{1-x}Ba_xSnS_4$ (CFBTS) alloys were studied with the SILAR method without any post-annealing process. When the partial substitution of Fe with Ba was conducted, the structural phase transition from stannite to trigonal phase was observed. The tunable band gaps were also achieved for the CFBTS alloyed thin films. Photoelectrochemical (PEC) cells were fabricated with the structure Glass/FTO/CFBTS with improved stability was fabricated without the addition of any bilayers on top of CFBTS thin films.

6.2 Future prospects

The following image depicts the future prospects of these quaternary chalcogenides and makes it further open to other applications as well **(Fig. 6.1)**.

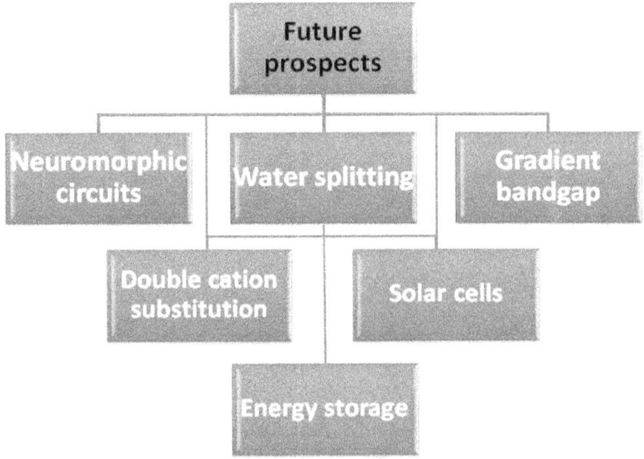

Figure 6.1 Future prospects of the quaternary chalcogenides.

As we have seen in our present work, the amorphous and crystalline phase affects the material band structure, hence the device resistance. Also, based on the different band structure and resistivity, the amorphous and crystalline phases can be distinguished simply by reading the electrical current. Such changes on chalcogenides are key enabling technologies for optical storage and also electrical non-volatile memory, known as phase-change memory (PCM). Non-volatile memory technology has been historically associated with charge trapping in flash memory. Hence, these CFTS based materials find wide application in oxide-based resistive-switching memory (RRAM). Resistive switching in RRAM is typically induced by bipolar switching, where the alternative application of positive and negative voltage pulses causes the formation/dissolution of a conductive (or insulating) region via ionic migration. These chalcogenides can be used in biologically inspired neuromorphic circuits because of their similar functions performed by the neurons and the synapses present inside the brain.

The neuromorphic circuits can achieve low power, highly parallel, and fault-tolerant systems which can have potential applications in the future. By fabricating neuromorphic hardware components, which can emulate functions performed by the neurons and the synapses present inside the brain. This phenomenon is not practical until the requirement of a large number of transistors is solved. So, to overcome these issues, hybrid neuromorphic architectures containing integrated neuron circuits with nanoscale device synapses fabricated with the help of chalcogenides can be used known as resistive memory devices. For that, an electrical model should be fabricated which suggests that for a device to emulate synaptic behaviour, it should have a conductance that can be precisely modulated by the type of stimulus it receives. Thus, the importance of material engineering of chalcogenides can be used for conventional memory applications such as multi-bit (multi-level) storage and the above proposed neuromorphic applications. We can also use these metal chalcogenides for implementing synaptic plasticity.

Other prospects of these metal chalcogenides are in the water splitting applications. Chalcogenides with the unique atomic arrangement and high electronic transport show interesting catalytic properties in various electrochemical reactions, such as the hydrogen evolution reaction (HER), oxygen evolution reaction (OER), and overall water splitting, while the control of their morphology and structure is of vital importance to their catalytic performance. A photoelectrochemical (PEC) tandem cell consisting of a p-type photocathode and an n-type photoanode, with the photovoltage provided by the two photoelectrodes, is an attractive route to achieve highly efficient water splitting at a low cost. These metal chalcogenides can cause photo corrosion under prolonged light irradiation also some of the metal oxides are corroded under bandgap excitation. So if we can prevent the photo corrosion and improve the photocatalyst properties then we can use it as an excellent photocatalyst.

To improve the charge separation and enhance the open-circuit voltage (Voc) in the absorber layers, the gradient-band gap materials need to be constructed by either solution-based or vacuum-based methods. Gradient band gaps have been mostly studied in the perovskite based materials which can be controllably turned to cover almost entire visible spectrum [167,168]. To the best of our knowledge gradient bandgaps have been rarely investigated in the Cu based chalcogenide devices. So a gradient band gap absorber film with a continuously tunable band gap can be demonstrated with the help of solution-processed methods. The gradient band gap device fabrication with the help of Cu based chalcogenides helps to improve the power conversion efficiency by reducing the recombination.

Double cation substitution in the CZTS absorber layer is an effective approach to suppress the cation disorder and reduce V_{oc} deficit. However, anti-site defects related to Cu and Zn form a midgap and deep level which significantly affect device performance. As low V_{oc} is considered as a major factor limiting the device performance, more research has to be done on the hetero-interface of kesterite solar cells. Fabrication of CFTS solar cells using a sputtering only approach without breaking a vacuum would be beneficial to reduce solar cells fabrication time and cost. Thin films of tunable band gap buffer layers are a possible candidate for the solar cell application. Also, careful control of the film composition along with required spike-like band alignment is needed to fabricate a highly efficient solar cell. Recently, metal chalcogenides are explored as transporting layers in the perovskite solar cells due to its better stability and efficient light-absorbing property. Also, we know that the bandgap of CFBTS thin films can be tailored from 1.5 eV to 1.8 eV. The ideal bandgap for better solar energy harvesting is 1.45 eV. This wider tunability in the bandgap value can be explored in several optoelectronic applications. Fabrication of solar cells on flexible and lightweight substrates is highly required to reduce installation costs and expand application areas. Therefore, more research needs to be performed to study these types of solar cells grown on flexible substrates such as foils or polymer-based substrates.

The expansion of technologies to convert solar energy and store it into a usable form of energy at large scale is of major importance worldwide. The integration of solar energy converting device with photo-chargeable supercapacitors as a single device has great potential to power various electronic devices. Different configurations can be used to integrate solar cells and storage devices, and the integration of these solar cells with supercapcitors can provide high-power conversion efficiencies. Introduction of more electrochemically active materials by increasing surface-to-volume ratio, can lead to favorable electrolyte ions diffusion and higher electrons transportation [169-173]. In addition, supercapacitors can play complementary role for energy storage solution, where batteries are not suitable for portable electronics, power back-up devices and other electronic applications. This type of integrated solar cell (energy harvesting) and supercapacitors (energy storage) share a common electrode. Thus, the direct integration of both energy conversion and storage components in a compact electrochemical cell is highly desirable.